George Henslow

The Theory of Evolution of living Things

And the Application of the principles of Evolution to Religion

George Henslow

The Theory of Evolution of living Things
And the Application of the principles of Evolution to Religion

ISBN/EAN: 9783337241063

Printed in Europe, USA, Canada, Australia, Japan

Cover: Foto ©ninafisch / pixelio.de

More available books at **www.hansebooks.com**

THE THEORY OF EVOLUTION OF LIVING THINGS

AND THE

APPLICATION OF THE PRINCIPLES OF EVOLUTION TO RELIGION

CONSIDERED AS ILLUSTRATIVE OF THE "WISDOM AND BENEFICENCE OF THE ALMIGHTY."

BY THE

REV. GEORGE HENSLOW, M.A., F.L.S., F.G.S.,
LECTURER ON BOTANY AT ST BARTHOLOMEW'S HOSPITAL, LONDON;
AUTHOR OF
GENESIS AND GEOLOGY, FREEDOM OF RELIGIOUS THOUGHT,
SCIENCE AND SCRIPTURE NOT ANTAGONISTIC, &c.

London:
MACMILLAN AND CO.
1873.

[All Rights reserved.]

This Treatise obtained one of the ACTONIAN PRIZES
for the year 1872.
5504 1

THE ACTON ENDOWMENT.

"HANNAH ACTON, of Euston Square, widow of Samuel Acton, architect, from motives of respect and regard for the memory of her late husband, and in order to carry into effect his desire and intention, and with a view to the diffusion and extension of useful knowledge in the Royal Institution, invested the sum of one thousand pounds sterling upon trust, so that at the end of every seven years successively one hundred and five pounds should be offered as a reward or Prize to the person, who shall in the judgment of the Committee of Managers for the time being of the said Institution, be the Author of the best essay illustrative of the Wisdom and Beneficence of the Almighty, in such department of Science as the said Committee of Managers shall in their discretion select."

[There having been no assignment of such Prize on the last septennial occasion—two Prizes were awarded in the year 1872, the other Essayist being Mr B. Thompson Lowne, F.R.C.S.]

PREFACE.

IN the following Essay I have abstained from offering any hypothesis whatever which might help to elucidate the methods of Evolution.

In the First Part I have endeavoured to give but little more than an outline, and that I fear a very imperfect one, of the evidence from palæontology; while one chapter only (VIII.) contains a summary from living organisms.

Being but indifferently acquainted with zoology and animal physiology, I am conscious of not having presented to the reader so convincing an argument as might possibly have been offered. Moreover, I have been compelled (in Chapter IV.) to assume him to have an acquaintance with much matter, which, unless he have studied palæontology, probably will not convey to his mind the same force as to the writer's. Still, if he rise from a perusal of this Essay with a stronger bias towards the Theory of Evolution of living things, and *will undertake to study nature impartially for himself*, my object will at least have been so far gained.

In the Second Part I have striven to shew that the very same laws of Evolution, which govern the origin and development of Beings, regulate as well the growth of Religion, whether in the Individual, the Church, or the Nation.

The argument is in fact cumulative, and its analogies are infinite. If the Reader can but realize this he will learn to regard the word 'Evolution' (not with suspicion or even horror!) but as expressing one of the grandest and most comprehensive laws in the universe.

It is unfortunate that some persons have entertained the idea that Evolution must be an atheistic theory. How it has arisen is not easy to say, unless it be from the fact that when a Man of Science advances a theory to account for certain phenomena in nature, he wisely omits all allusion to any theological questions—not that he may not, or does not, the while hold views of Divine action in his own mind. Thus probably has it been with the Theory of Evolution. This word has been adopted simply to express the conviction forced upon the minds of certain observers, that all organisms are derived from preexisting ones, and not that each kind has been separately created. Moreover, it is not that nature is called in by the theorist to support his views—as some seem to imagine—but that Evolution is best able to express most of, if not all the phenomena connected with the origin of beings.

That the Theory of Evolution is not necessarily Atheistic, I trust this Essay will clearly shew[1].

---

[1] Of course, I am aware that the Theory of Evolution, as held by some persons, may be in appearance, if not actually atheistic; but all views of Evolution are not to be condemned, because materialists and positivists may profess to dispense with the aid of a Deity in Creation.

Perhaps it may be worth while to add, that just as it is believed a man's spiritual growth in grace to be due to the divine aid of the Holy Spirit, working through and with his Free-Will, to enable him to "work out his own salvation": so, I believe, does God in nature by means of powers or laws,—the nature and method of action of which are quite unknown to us,—work out His own purposes through Evolution, and produce those effects, which have given the impression that they were directly designed.

In speaking of design I would observe, that the so-called 'argument of design' is not at all weakened, but rather strengthened by the Theory of Evolution: for we need but extend Paley's remarks upon the watch to see this. He alludes to the mechanism as proof of an intelligent maker, and adds that our idea of the greatness of that intelligence would be much enhanced, if watches could produce offspring like themselves. Now, surely, *a fortiori* would the intelligence appear far greater, if the maker could have infused into the watch, not only reparative powers in case of injury but also the law that slight variations should appear in the offspring, which, by accumulating in successive generations, ultimately produced all the varieties not only of watches but also of clocks of every description in the world. Yet this would be exactly analogous to the production of animal and vegetal organisms by Evolution.

The *design* seen in a structure, is in fact part of, and inseparable from it, pointing at once to the existence of Mind, and is quite irrespective of the *process* by which that structure was brought into existence.

P. S. I have accidentally omitted an acknowledgment of the aid I have derived from Mr H. G. Lewes' papers in *Fraser* and the *Fortnightly* for some information embodied in the second Chapter.

CONTENTS.

PART I.

THE THEORY OF EVOLUTION OF LIVING THINGS.

CHAPTER I.
INTRODUCTION . . . . PAGE 1

CHAPTER II.
HISTORY OF EVOLUTION 15

CHAPTER III.
THE IMPERFECTIONS OF THE GEOLOGICAL RECORD . 32

CHAPTER IV.
GEOLOGICAL SUCCESSION OF VERTEBRATE LIFE . . 41

CHAPTER V.
INVERTEBRATE TYPES 77

CHAPTER VI.
NEGATIVE EVIDENCE 84

CHAPTER VII.
HOMOTAXIAL LIFE 91

CHAPTER VIII.
SUMMARY OF THE EVIDENCE OF EVOLUTION FROM EXISTING BEINGS 97

PART II.

EVOLUTION AND RELIGION.

CHAPTER IX.
	PAGE
MAN	107

CHAPTER X.
THE ARGUMENT FROM ANALOGY 124

CHAPTER XI.
MORAL EVIL OR SELFISHNESS 143

CHAPTER XII.
ON THE WISDOM AND BENEFICENCE OF THE CREATOR 158

CHAPTER XIII.
THE LAW OF INIDEALITY AND THE IDEA OF PERFECTION:—CONCLUSION . , ... 197

PART I.

THE THEORY OF EVOLUTION OF LIVING THINGS.

CHAPTER I.

Introduction.

THE theory, or rather doctrine, of the Evolution of Living Things has not yet received that uniform acceptance to which it is undoubtedly entitled. That it will in time become generally received may be reasonably presumed; but at present, with many theologians at least, the Creative hypothesis obstinately holds its ground. Two causes may be assigned to account for this fact. First, there is the preconceived but erroneous idea of the method of creation derived from a misconception of the first chapter of Genesis. Secondly, there is the unfortunate but very general want of any scientific training, not only amongst the clergy but in the public generally; and, as a result, there is that absence of a due power of appreciation of the arguments of the scientific man, which is so conspicuous in their style of reasoning.

In order, therefore, that the proof of the Wisdom and Beneficence of the Almighty as shewn in the processes of Evolution may not be considered as based on unsound premises, it will be desirable to point out the untenableness of the present theological position, as well as the grounds upon which Evolution is founded; and which will, let us hope, be soon recognised as incontrovertible by all who seek the truth in earnest. Until comparatively recent times the Book of Genesis was supposed to reveal in its first chapter an explicit account of the origin of Living Things, namely, by direct Creative Fiats of the Almighty. All the known animals and plants being far fewer than at the present day, their differences were more pronounced than their resemblances. Each animal and plant was observed to bring forth its offspring "after his kind," generation after generation, without any noticeable change. Any other animals than those now living on the globe were never conceived. Fossil shells were supposed to be either deep-sea creatures thrown up upon the beach, or, if found on land and upon hills, easily accounted for by the Deluge.

Every living thing was believed to have been created at once by the Word of the Lord: and all within the space of six literal days.

When Geology came to be studied with some philosophic spirit, it was soon discovered that many fossils were not of living species; that six days was incontestably too short a period to account for geological phenomena; that a flood, even if conceded to have been universal, was unable to solve many a problem of disturbance and stratification. Moreover it was perceived that the Earth's structure was separable into several strata; and that each stratum contained a group of fossils unknown either in the stratum above or below it; and upon this discovery was based the principle, that disconnected strata might be recognised by the identity of their organic remains. In addition to these facts the phenomena now known as *dislocation, contortion, upheaval, unconformability* and others frequently occurred, and apparently often during periods intervening between the deposition of strata.

These latter appearances, taken into consideration with the daily phenomena of Volcanic Action, induced the Geologist to conceive, and the Theologian to adopt, the theory of successive creations after cataclysmic and predetermined destructions of all existing life by the Almighty: while to meet the now well-established truth of almost infinite ages having elapsed, the Theologian adopted the interpretation of *Ages* for the Hebrew word *Yōm* or *Day*. If, however, the first chapter of Genesis be read without any reference to or thought of geological discoveries, and the first three verses of the second chapter be carefully compared with the fourth Commandment, it will not appear how any notion of an indefinite time can be given to the word 'Day' at all. The writer of Genesis (as I have expressed elsewhere[1]) seems to signify a Day in the ordinary sense and apparently without any conception of indefinite periods at all.

Geology ceased not to pursue her avocations steadily and uncompromisingly.

The study of the rocks soon brought to light a large increase of the number of strata: so that at the present day there are *thirteen* 'formations,' embracing *thirty-nine* principal 'strata,' the strata themselves being often subdivided into minor ones. If therefore the miraculous re-creations be true they must have been very numerous. But with the discovery of additional strata a larger insight was obtained into the distribution in time of animal and vegetal life. It was then discovered that these 'created groups' were not so rigidly defined as at first supposed, and consequently the rule established by Geologists themselves can only be applied

---

[1] *Genesis and Geology, A plea for the Doctrine of Evolution. A Sermon:* Hardwicke; in which the idea of *six Visions* is reservedly suggested. Cf. *Testimony of the Rocks*, by Hugh Miller.

cautiously in attempting to parallel distant strata:—though some species appear to characterize strata respectively, yet many range up and down through other than those in which they attain their maximum development or of which they may be especially characteristic.

Two difficulties thus arose:—the increase of miraculous interferences *seemed to increase proportionately their improbability;* especially as there was no corroboration this time from the Word of God; while the fact of species ranging through several *strata* threw another stumbling-block in the way of the cataclysmic theory; for either they must have been re-created two or three times, or else lived through the supposed cataclysms considered as designed methods of destruction.

Another class of phenomena now appeared, to shew a still greater difficulty in the way of belief in the Creative hypothesis. Zoology, Botany, as well as Palæontology, gradually increased the number of living and extinct forms almost indefinitely; and in proportion as fresh discoveries were made, so it was found that numbers of forms took up positions, when classified, *intermediate* to other forms hitherto well distinct—'osculant' or intercalary forms as they are called. These often increased so much, that even genera well marked at first became blended together by transitional or intermediate forms.

Hence it has come to pass, from the result of this discovery, that so far from Forms or Types of organisms being easy and of a precise character, in accordance with the idea of each being well defined after his kind, systematic Zoology and Botany are the most difficult tasks a naturalist can undertake. Here, then, an overwhelming difficulty, only to be fully appreciated by a really scientific person, rises against the conception of each kind having been specially created as we see them now. Indeed it may be added that the very idea of *kind* or *species* has been resolved into an abstract conception finding in Nature generally no more than a relative existence.

Fresh difficulties were still in store, which must be overcome if the former theory of Creation is to obtain any longer—Horticulture, Floriculture, Agriculture, and the Breeding of Animals have rapidly risen to become important and flourishing occupations. From their pursuit it was soon discovered that kinds reproducing their like *never did so absolutely*, but that offspring appeared always to differ from their parents in some trifling if not considerable degree. This property of Nature, to which also the human race is invariably subject, Man has seized upon, and by judicious treatment can almost mould his cattle to whatever form he pleases, or stock his fields and gardens with roots of any form or with flowers of any shade of colour required. After many years of successful propagation, generation after generation, we have now arrived at the result that animals and plants can be produced by careful breeding and selection, which, had they been wild, our earlier naturalists would have undoubtedly regarded as having been respectively created at the beginning of the world! Here then we have a practical basis of argument to account for the many transitional forms which Geology reveals in the past history of the world, as well as amongst the plants and animals living at the present day.

Yet another fact may be mentioned. Geographical Botany and Zoology began to be studied as travellers stocked our Museums and Herbaria with an ever-increasing number of beings brought from all parts of the world; and the, so to say, capricious distribution of identical forms in far distant places—now explicable on the theory of migration and subsequent isolation—as well as the appearance of representative forms of allied though different kinds in certain districts, explicable *only* on the theory of descent with modification, have a strong *prima facie* appearance against the theory of individual creations, even if Geology did not furnish undoubted evidence of very frequent interchanges between land and sea having taken place. Without at present giving more reasons, the above will be sufficient to shew cause why

Science has found herself compelled to secede from the cramping toils of the Creative hypothesis, and to take up that of the Evolution of Living Things as better explaining *all* the foregoing phenomena. In proportion as the probability of the former was seen to decrease, so in the same degree does that of Evolution increase. Hence, at the present day the argument in favour of development of species by natural laws may be stated in the following terms, viz., 'It is infinitely more probable that all living and extinct beings have been developed or evolved by natural laws of generation from pre-existing forms, than that they with all their innumerable races and varieties should owe their existences severally to Creative Fiats.'

But even now asks the Theologian, Does not this theory controvert the Bible, for we are distinctly told that God created everything after its kind?

In reply it may be confidently shewn that the Theologian cannot be sure of the value of his interpretation of the first chapter of Genesis, at least so far as he attempts to draw scientific deductions from it. Thus it may be observed to him that the words 'create' and 'make' are used indifferently; that no definition is given to ensure accuracy as to their right interpretation. It is not stated whether God created out of nothing or out of eternally or at least pre-existing matter. Moreover, in addition to the statement that God created or made all things, there is the oft-repeated assertion embodied in the word *Fiat*, but apparently overlooked, that He enjoined the Earth and the Waters to bring forth living forms. What does this expression imply?

The use of the imperative mood can only signify *an agent other than the speaker*. If therefore it be maintained that the sentence (ver. 21) "God created every living thing that moveth" signifies He made them by His direct Almighty fiat, it may be equally maintained that the sentence "Let the waters bring forth abundantly every moving creature" implies secondary agents to carry out the will of the Lord. Such might be said to witness to natural law, which, after all, is but a synonym for the will of God.

The real basis of the controversy between Dogmatic Theology and this deduction of Science is simply this:—The former has established a creed based upon erroneous impressions derived from Scripture, and, from having had power in former days to enforce its opinions, they were credulously received without hesitation as long as no one dared to or even could controvert them. It is the reluctance to surrender this power to Science as much as the idea of her offering any opposition to Theology that urges at least one body so obstinately to resist her advances. Nearer home the opposition rests more on the latter ground; and it will not be until the representatives of our Theology can see and confess their false impressions of the meaning of the first chapter of Genesis, that the doctrine of Evolution can be hoped to make any great progress amongst them.

Let us briefly review their false positions. They first clung to the 'six days of Creation;' they found they were compelled to surrender the idea and immediately adopted the interpretation of *Yōm* signifying an indefinite period. Again, notice their readiness in adopting the theory of cataclysms and re-creations, a second time to the detriment of Genesis, which furnishes no warrant for the idea; for even if six days be presumed to represent six cataclysms, Geology furnishes no corresponding evidence. It was a pure fiction altogether. And even now they steadily oppose the doctrine of Evolution. But surely as each stronghold of Theology has been quietly taken by Science—not so much by offensive attack as by undermining and leaving the edifice to crumble of itself—the tardy and ungracious capitulations hitherto offered only ensure the ultimate surrender a matter of patient expectation. *A time will shortly come* when the Creative theory must succumb altogether, and the doctrine (not the theory) of Evolution will be as much recognised as a fundamental truth of Science and Theology as the revolution of the Earth itself.

It may be advisable to state that, in the following essay, nothing will be said as to the first origin of life on this Earth; for no evidence is forthcoming from Nature, upon which even the merest random speculation could be founded[1]. Similarly it seems advisable to leave any question of the method of development of Man from the lower orders untouched; though some contrasts will be drawn between them.

---

[1] I shall also make no allusion elsewhere to Spontaneous Generation; for whether it be true or not, it affords no direct or indirect aid in maintaining the Theory of Evolution.

CHAPTER II.

HISTORY OF EVOLUTION.

IF we endeavour to trace the history of the doctrine of Evolution, it does not appear to afford much definite or scientific aspect until the time of Wolff, who promulgated the theory of *Epigenesis* in 1759: though while doing justice to him as the real precursor of the modern expositors of Evolution, we must not forget the names of Robinet, Bonnet, Geoffroy S. Hilaire, Meckel, Lamarck, or the author of the *Vestiges of Creation*, and lastly Mr Darwin, Mr Wallace and other living authors, who have helped to found the doctrine on a more scientific and sounder basis. The writers of the last century were for the most part too deeply imbued with metaphysical ideas of subjective philosophy, while objective science was too little studied in the exact spirit of modern days, to allow of any accurate or even rational theory which could meet with much or even any acceptance now. It may not be unadvisable to give an instance of the gross absurdities which in former days were allowed to pass under the name of science.

In 1766 a work entitled *De la Nature* was published at Amsterdam, written by T. B. Robinet. It rapidly went through three editions, which proves how popular it must have been. In this remarkable mixture of pseudo-natural science and metaphysics the author proposes to accept the idea *in toto* of a 'chain' of beings. By this he understands that all things live, earth, rocks, and even stars, as well as plants and animals. All things alike have sentient perceptions, all grow, desire and propagate. He recognises fully under the idea of desires the two properties of *Growth* and *Reproduction* as qualities of all things. Thus, he says, fire is hungry and voracious; it feeds on air, and if air be wanting, it expires. So, too, air feeds on water, while water feeds on other substances:

and thus he accounts for such minerals as salt, iron, &c. being found in mineral springs. The following quotation will illustrate his style of reasoning:—

"I have sought for the germs of stones and the vessels which contain them; nor have my researches been fruitless, I have even discovered how stones and minerals eject their germs. If I have been unable to detect their sexes, how many animals and plants are in the same condition. Finally we have seen an infinity of fœtal stones and metals in their wombs, with their envelopes and placentas; we have seen them growing and nourishing them, like animals. There may be stones which multiply by budding, as is the case with trees and some animals. But observations are wanting to confirm this conjecture."

Thus believing, Robinet asserts all things to be organic; and just as he would recognise a crystal to be composed of molecular crystals, each having the form and properties of the whole, so an animal is composed of a vast number of minute animals, a dog of little dogs, a man of molecular men! As an illustration more apposite to our theory Robinet traces the transitions from a tadpole to a frog; adding, however, that not only is this an instance of change of one creature into another, but what has certainly no affinity to the modern doctrine of Evolution, that frogs can change back into fishes! for thus he says:—"It seems that the frogs of all countries are fish or tadpoles before they become frogs. It is not equally asserted that in all countries the frogs change again into fish, as is the case with those of Surinam, Curaçoa, and other parts of America. This singular metamorphosis of a little fish changing into a frog and a frog back again into a fish offers a vast field to the reflections of the naturalist."

These quotations will suffice to shew how inaccurate must have been the observations, and how crude the ideas, of certain men of science a century ago.

Another instance of quite as untenable an hypothesis may be mentioned.

Observations upon the phenomena of descent, such as the form, features, dispositions and diseases of parents and grand-parents being often reproduced in descendants, each of which takes its origin in a microscopic atom or cell, gave rise to the theory of *Emboîtement* or preformation, *i.e.* that the original germ of every species contained within it all the countless individuals which in process of time might issue from it, 'boxed up' as it were potentially.

This theory found many adherents; even Cuvier himself appears to have been amongst them. But it was the fate of this, as of all previous theories, to submit to the supremacy of another; and the one which completely put aside the emboîtement theory was that of Epigenesis by Wolff, who, in the same year as Robinet, propounded it in his *Theoria Generationis*. This may be justly considered the true starting-point of the history of Evolution, at least in that form in which it now finds general acceptance as a genuine scientific doctrine.

Wolff endeavoured to express the fact that,

instead of each organism being as it were included *in toto* in previous ones, each organ arose in succession upon the being or organism, having been secreted from or deposited on another. The general result of the theory is that every part of an animal or plant is the *effect* of a pre-existing part, and this in its turn furnishes the *cause* of a new and succeeding part.

It will be at once conjectured from this that the law of Epigenesis implies a necessary process of serial development, and that all organs issue in accordance with the impulses of the inherent momenta of the individual.

We must notice that the idea of preconceived types or plans as having entered the Mind of the Creator is precluded from this consideration or belief altogether.

We need not discuss further the merits or demerits of a theory which scarcely meets with any approbation at the present day, though it cannot be denied that it closely approximates to the general theory of Evolution, especially to those forms which are held by Positivists and others who do not recognise 'plans or types' as indicative of a Designer or God.

The next theory worthy of notice was that of Lamarck, which he produced in 1809, and which, like its predecessor, had a brief day, until it was revived 50 years later under the auspices of the *Origin of Species*.

Lamarck based his theory upon the observation that organs become hypertrophied by use and atrophied by disuse; that a change of external conditions produces analogous changes upon individuals, both animals and plants.

Now every change of environment brings forward a new want which must be met by the individual if it is to retain its ground and propagate: and these new wants urge the being to new actions and new habits. Thus organs, not much used under previous conditions, obtain importance, are exercised, and develop into new organs. On the other hand, organs now being disused become atrophied and dwarfed, or disappear altogether.

Assuming, for what, by the way, there was

not a shadow of foundation, that an alteration of external conditions about a being may cause certain organs to disappear while new ones are developed, Lamarck enunciated the proposition that it is not the organs of an animal which determine its habits; but its habits which determine the form of the body. For example, the Antelope and Gazelle were not specially endowed with agile limbs *in order* to escape from their enemies, but having been exposed to the danger of being devoured, they were compelled to exert themselves in running. This habit then gave rise not only to the extreme agility so characteristic of them, but actually to the slender limbs themselves as well, upon which their swiftness of running depends.

According to this theory animals and plants have advanced by a graduated scale of a less to a more complex state as they were successively developed during the past ages of the world: a priority being given to marine types over terrestrial ones, which were the improved forms of such aquatic kinds.

It will be observed that this theory is directly opposed to the philosophy of the ancients, who imagined a golden era to have prevailed during the first condition of things and that all sublunary matters when left to themselves tend to degenerate. Whereas, according to Lamarck's and subsequent phases of the doctrine of Evolution, incipient and imperfect conditions obtained, which have in turn been replaced by more specialised and perfect forms.

A principle of progressive improvement was presumed inherent in all substances, so that inert matter first acquired life, then sensation was developed out of vegetative vitality. Instinct and mental faculties followed; and at last by the same processes what had been irrational became rational.

Now, in order to overcome the difficulty of the fact of myriads of creatures of all grades still existing, and shewing no sign of or tendency towards 'improvement,' Lamarck conceived Nature, which he recognises as a mere instrument and obliged to act according to laws impressed upon it by God, to be continually giving rise to elementary rudiments of animal and vegetable existence by spontaneous generation: *Monads* according to his theory are being produced every day, which in course of ages will be developed into beings higher and higher in the scale of life.

Such, briefly, are the principles of Lamarck's theory, the final result being the supposition that Man has been developed out of the Ourang-outang. The idea of spontaneous generation being a fact was clearly a mere hypothesis on the part of Lamarck in order to account for the infusoria and other low types still existing. He could not adduce any more proof than Lucretius of old, who indeed was somewhat bolder in his assertion, that Nature bears natural wombs attached by fibres to the surface of the ground, which give rise to animals!

It is curious to find how subsequent ages have revived this idea, though now based upon more scientific if not very accurate experiments.

The belief has held sway and does still over many minds, though it is probably due to Professor Huxley that an end has been put to the controversy, let us hope, for ever: so that once more spontaneous development is relegated to the limbo of exploded theories.

On the other hand 'Persistence of Type' is the expression now adopted to signify that probably of almost every group that has ever existed on the Earth, at least some forms of relatively lower grade still survive; so that examples are known of organisms having undergone in their descendants little or no change since the earliest known ages of the history of life[1]. Lamarck's theory of development as well as its subsequent exposition by the author of the *Vestiges* have both in turn succumbed to that of Mr Darwin, who 50 years after Lamarck propounded the theory of 'Natural Selection' as the principal operation by which Nature evolves new species. This theory is too well known to call for any lengthened examination in this essay. It will be sufficient just to point out the facts upon which it is based.

[1] Such, for example, as the *Lingulella* of the lower Silurian strata, as well as many of the *Foraminifera*.

It is an established fact that all animals and plants as a rule produce far more offspring than can possibly ever come to perfection[1]. The lower the grade, the greater is the progeny. Hence is induced an intense 'struggle for existence.' If therefore any, by possessing a more vigorous constitution or what not, can gain an ascendancy over the rest of the offspring, it will live while others will die. Now it is an obvious fact, that while parents in the main produce offspring like themselves, that is all have a general family likeness, yet these differ from their parents in certain elements of their structure.

Those, then, which are 'best fitted,' survive. These variations of form which some of the offspring exhibit being beneficial become hereditary and intensified, as well as give rise to other changes of structure of correlative development: the general result being that changes in organic forms arise to keep organisms in

[1] From a calculation made by the writer, he found *one* plant of the Foxglove yielded, in one year, more than a million and a half good seeds. This is far less a number than would be in the case of the spores of agarics.

harmony with or in adjustment to external conditions, and thus, gradually through successive generations, new forms arise differing so much from their original parent, or parents, that they are pronounced to be new species.

It cannot be too strongly borne in mind that all these writers are not only witnesses to the deeply-rooted belief in Evolution generally, but are theorists who endeavour to account for it by some law of Nature.

The object of the present essay is not to propound any new theory, nor to express direct approbation of any or all of these theorists' opinions: but simply to shew what are the evidences for the doctrine of Evolution of Living Things, and how the wisdom and beneficence of God are shewn in the phenomena of creation as based upon that doctrine.

It must be noted, too, how all of those theories have in turn fallen into abeyance except that of Mr Darwin; and there are not wanting many symptoms of decay in the acceptance even of his. Not only has he considerably modified his views in later editions of the *Origin of Species*, distinctly expressing his opinion that he attributed too great an influence to natural selection, but even men of science such as Owen, Huxley, and, at least in its application to Man, Wallace himself, are either opposed to it in great measure or else give it but a qualified assent. Thus, it has been the fate of all theories of the development of Living Things to lapse more or less into oblivion. *Evolution* itself, however, will stand the same. This is the great fact of Nature, while those theories to which allusion has been made are but suggestions to account for it. Each theorist has caught hold of some one fact in Nature and held it up as the sole cause of Evolution, which is without doubt the resultant of many causes, no one alone being sufficient to bring it about, though each has been ever playing its part in the grand scheme of Development.

There is unfortunately great confusion in the public mind on this subject; for many persons seem to think that 'Darwinism,' the 'Development hypothesis,' and 'Evolution,' are expressions of equivalent value; whereas the first two are theories erected upon certain facts existing in Nature in order to account for the third or the general principle of Evolution.

To define, therefore, clearly what is to be understood by the term 'Evolution' is highly desirable. Briefly, then, it supposes all animals and plants that exist now or have ever existed to have been produced through laws of generation from pre-existing animals and plants respectively; that affinity amongst organic beings implies, or is due to, community of descent; and that the degree of affinity between organisms is in proportion to their nearness of generation, or, at least, to the persistence of common characters; they being the products of originally the same parentage.

It must be borne in mind that evolution or differentiation is not, so to say, *necessary* in every organism, for in many cases little or no differentiation has taken place. Indeed were it not for this fact the world would not now have representatives of any group except the highest, (unless Lamarck's idea of constantly-arising monads to give birth to new successions of beings be adopted!) whereas it is abundantly supplied with members of every kind from the humble rhizopod up to man.

This retention of character without differentiation has been called by Professor Huxley *Persistence of type*, and probably covers a larger area of the organic world than even Evolution itself!

As nothing is known of the origin of life, or what the first creatures were, or indeed how life arises at all, it is presumable that God was the Author of it; but that assuming Him to have once created a being or beings endowed with life, Evolution has produced all subsequent ones in accordance with the axiom 'Life only can proceed from Life.' That is to say the methods of Deity in peopling this world with animals and plants throughout its successive ages until now, has been called *Evolution* by men of science. It would be out of place to enter into any discussion of the views of the Positivist or Atheist, who ignore or deny the existence of any evidence of a Creator. The writer of the present essay assumes God to be the Author of Creation and believes Him to have adopted Evolution as the method by which He chose to bring about the existence of successive orders of beings until Man appeared upon the scene of Life. At the same time he does not venture to pronounce as to the secondary causes or instruments by which this has been effected.

CHAPTER III.

THE IMPERFECTIONS OF THE GEOLOGICAL RECORD.

IN tracing the history of life upon this world it is obviously impossible to acquire any adequate ideas of the origin of living things, if we limit our observations to such only as have existed within the history of man. To secure anything like a complete review of life we must examine every trace of its existence, possible to be obtained, during the past ages of the world: and argue from them as to the most reasonable or probable account of their origin and existence.

In attempting this the first and obvious fact which strikes the observation is the extreme imperfection of the geological record: so that before demonstrating what facts of Geology may be considered worthy of note as supporting the doctrine of Evolution, it will be very advisable to preface the investigation with a distinct exposition of the several grounds for recognising the great imperfection of the palæontological records.

Thus there exists the great law of dissolution of the 'organic' into the 'inorganic,' with the consequent loss of the organism.

Now it is obvious that great differences obtain in the extent of this dissolution. All soft parts in general rapidly disappear; and hard parts, such as bones, teeth and shells, though more durable, can only be preserved under favourable circumstances; consequently numerous and whole groups of animals and plants have possibly never left any record of their existence at all; such, for example, as the *Tunicata, Holothurida* and many *Hydrozoa*.

Again, dead animals are preyed upon by others; and we may notice how rare it is to find the bones of birds, mammals, &c. now; while ordinary experience proves how little likely animals are to become buried in lacustrine or other deposits.

We may record here an observation by Mr Darwin: that a large proportion of bones of mammals of the tertiary period are found in caves and lacustrine deposits; while not a single cave or certainly true lacustrine deposit is known in the mesozoic or palæozoic strata.

If, therefore, it be a rare thing to find the dead bodies or rather skeletons of any of the millions of vertebrates that die annually, how greatly is the chance of discovering the few that happen to have been immured, during past geological ages, decreased, and how much less must this chance be when we investigate the earlier strata of the earth!

The following considerations will shew this:

(1) Any particular stratum is the exact measure of the denudation of some previously existing rocks, and indicates so much loss of the previous history of life which would have been represented by the fossils of those lost rocks.

And, as strata can be measured by miles in thickness, the corresponding destructions have been just as enormous[1]: not to add that the materials of any stratum may have been deposited with organic remains in them, then destroyed and subsequently redeposited with a new set of organisms over and over again.

(2) Supposing organic remains to have been preserved in deposits, to what extent has the earth been explored by geologists? The portion of land carefully searched bears but an excessively small proportion to the extent not yet examined: while the sea bottom, *i.e.* three-fourths of the entire surface of the globe, cannot be searched at all. So that it is really somewhere about the ten-thousandth part altogether which has been explored.

(3) To have a perfect series, the successive strata of the whole surface should be searched

---

[1] See this further discussed by Mr Darwin in the *Origin of Species*, 4th ed. p. 344; though he does not observe that not only is any particular stratum the exact measure of a previously existing denudation, but it is only *the last instance of denudation* of that particular amount of material.

for each period of the world's history; as we may reasonably suppose that the entire world was peopled with creatures successively, as age after age followed each other. Hence, we really want a complete series of fossils not only from each stratum in one locality—say Great Britain—but from all parts of the world.

(4) The great destruction of old land surfaces, with all the varied forms of terrestrial and aërial life, causes many a hiatus which cannot be refilled: so that an enormous lapse of time may have occurred between the deposition of two conformable strata, of which the lower had furnished for ages a land surface, but which subsequently became denuded by marine action, and then was the recipient of a new deposit, and so now bears upon its upper surface no trace of any evidence of the ages that had intervened between the depositions of the two strata.

In vain may we seek for the old land surfaces upon which the Mammalia roamed which gave rise to the Eocene *Lophiodon, Coryphodon, Anoplotheria*, and *Palæotheria*. Again, where was the land upon which the reptiles lived which were the forerunners of the secondary Dinosauria? where are the ancestors of the Marsupials of the Oolites? or again, the land which bore the antecedents of the Devonian Conifers? Not until we come to the tertiary epoch and especially the middle and later periods do we find a great abundance of terrestrial life: and it may be observed it is just this very abundance which furnishes the most valuable evidence in favour of the doctrine of Evolution: but these remains have been preserved in caves and lacustrine deposits; such being, as we have remarked, unknown in strata antecedent to those of the tertiary age.

The old idea of the accidental discovery of a new form being indicative of the creature's first appearance on the earth is absurd and utterly inadmissible. On the contrary, the earliest observed presence of any well-defined form is in itself a witness to lost, and possibly a long line of lost, ancestors.

Similarly we must be cautious in asserting that because a certain form disappears from a certain stratum which follows one in which it had been abundant, that it then vanished from the world altogether. The part which *Migration* has played in distributing forms of life is of immense importance in the consideration of the history of life. It has been too much ignored by far; and it is only by correcting our ideas of the development of ancient life by the study of the existing geographical distributions of animals and plants, that we shall guard ourselves against any hasty generalizations.

Mr Darwin enters somewhat at length into the discussion of the absence of intermediate forms or varieties which should connect different species of fossils; and he believes that their absence is due to the great imperfection of the geological record. According to his theory there ought to have existed many gradations between any two forms well differentiated in time by descent. Now it may be essential to the support of the theory of 'Natural Selection' to believe such to have existed, but it is not essential to the truth or acceptance of the general doctrine of *Evolution*. Quite enough of transitional or intercalary forms have existed, and more than enough of living forms are to be found, to support this doctrine; or, to speak more strictly, so many exist as to force themselves upon the attention of the biologist, who finds no more satisfactory explanation of them than the doctrine of Evolution. Though it often happens that *minute* gradations are wanting, there are indeed many grounds for thinking Nature occasionally, perhaps often, proceeds by 'leaps' or 'sports' as well as almost insensible gradations in the evolution or development of species.

These considerations will prepare the reader not to expect very strong or even satisfactory evidence from Palæontology, more especially from the records of palæozoic or mesozoic times. Indeed, so fragmentary is the record, that our greatest biologists *have* hesitated to bring forward evidence in favour of Evolution from either of those epochs at all; though, as will be seen hereafter, perhaps with undue timidity.

On the other hand, opponents of the doctrine of Evolution invariably take their stand upon this, so to say, abandoned outpost of palæontological ground, ignoring the existence of many kinds of evidence to be deduced not only from tertiary Mammalia but from living organisms as well.

If now we admit thus fully that all geological observation of the history of life has been necessarily so extremely scanty, then whatever evidence it may afford in favour of Evolution will be of proportionally greater value.

CHAPTER IV.

GEOLOGICAL SUCCESSION OF VERTEBRATE LIFE.

FROM the considerations given in the preceding chapter, it is obvious that to expect good evidence in favour of Evolution we must search those places where the above-mentioned causes of imperfection of the geological records least occur.

First, then, it is clear that the later the period, *cæteris paribus*, the better will be the chance. The tertiary period will, therefore, be preferable to the secondary, or primary, in consequence of there having been less destructive denudation of those kinds of deposits (especially lacustrine) most favourable for the preservation of the relics of terrestrial life.

Now, as a matter of fact, whether they have been subject to much denudation or not, the

tertiary strata *do* furnish us with the materials we want, the Miocene and subsequent periods especially having left an abundant array of fossil Mammalia for investigation.

Now, the nature of the evidence supplied by Palæontology and especially by the fossil Mammalia of the tertiary epoch may be arranged conveniently under the following heads.

1. *Structure* or *Form.* The existence of 'more generalized,' 'average,' or 'less specialized' forms, *i.e.* when compared with others of the groups or types, is now established. The following are good illustrations ;—*Anchitherium, Oreodon, Ictitherium, Zeuglodon, Dinotherium,* &c.; while *Compsognatha* of the Dinosaurians and *Ichthyosaurus* represent the same features amongst the Reptilia of the Mesozoic ages.

2. *Intercalary* or *intermediate forms* are very abundant; perhaps more frequently to be found amongst the Mollusca, simply because large suites of fossils can be more readily obtained: but of the Vertebrata some of those mentioned above might be regarded as 'intercalary' as well

as being much generalized. Thus *e.g. Zeuglodon*, while bearing affinities to the *Sirenia*, is more strictly intercalary between Seals and Cetaceans. Again, there were forms called *Stegodon*, linking *Mastodon* with the Elephant; while the genus *Hemiarctos* united bears and dogs.

3. *Embryonic types*, and those of relatively low grades, are not wanting. Thus *Archegosaurus*, of the Carboniferous age, is at least embryonic in several features, while the Labyrinthodonts were not of the highest grade of Amphibians. Many of the palæozoic fishes with the persistent notochord are other instances, while the unankylosed condition of the vertebræ of Ichthyosaurus and that of the epiphyses of the femur of Plesiosaurus, &c. all point to the same phenomenon, *i.e.* a permanently arrested condition of development, or permanency in the adult of embryonic features.

4. On the other hand, *well differentiated forms* and highly specialized structures have existed in all ages of the world that have furnished strata for examination.

5. *More generalized faunas* existed in certain localities, where more specialized ones are now found; thus the Miocene fauna of Europe contained Apes, Gazelles, Edentates, and Marsupials; types now widely dispersed. Similarly it has been shewn that the Miocene fauna of North America abounded in forms of more generalized types than the European, while wanting in many forms abundant in the Old World at that age.

6. *Local faunas* date back to at least Post-pliocene times, and are doubtless the modern representatives by lineal 'descent with modification' of similar faunas though of extinct forms; for example, there are the Edentata of South America, the Marsupials of Australia, the mixed fauna of India, which may be compared with a similar Miocene group of the Sewâlik hills, and the Struthian birds of the Southern islands.

7. *Migration* of types of local faunas into other countries has taken place, as the Miocene European forms are met with in Pliocene American deposits.

8. *Homotaxial*[1] *life* has ever existed in synchronous strata generally throughout the world in all ages. The more ancient the strata, the nearer is the identity of fossils maintained in separate strata of the same epoch: as *e.g.* in the carboniferous limestone of Europe and South America.

Before, however, attempting to apply these principles of evidence of the probability of Evolution being the true theory, it will be as well to state generally what are the facts revealed by geological discoveries as to the appearance of different groups of animal life successively upon the face of the globe.

It would be out of place in an essay like the present to enumerate more of the various kinds of animals and plants discovered in different geological strata, than may be necessary for my purpose of giving an outline only of palæontological evidence of Evolution.

---

[1] This signifies that "certain forms of life in one locality occur in the same general order of succession as similar forms in another locality." 'Anniversary Address' by Prof. Huxley, *Quarterly Journal of the Geological Society*, Vol. XXVI. p. xlii.

Reference to any work on Geology will give the evidence upon which the following facts are stated.

Confining our attention for the present to the Vertebrata alone, the remains of Fish only have as yet been found in strata older than the Carboniferous; and so characteristic were certain types of fish during the middle and upper palæozoic epochs as to give rise to the expression 'Age of Fishes' as applicable to those periods. As illustrations of such types may be mentioned the remarkable forms, *Pterichthys*, *Coccosteus*, and other ganoids: *e.g.* the Crossopterygidæ (of Huxley).

Secondly, no vertebrate of higher organization than an Amphibian has ever yet been found below the Permian strata: while, of the extinct Amphibia, there were not only numerous forms but many of them gigantic in size, *e.g.* the *Labyrinthodontia*, thus entitling that age of the world, *viz.* from the Carboniferous to Trias inclusive, to be called 'the Age of Amphibia.'

Again, from the Permian to the Cretaceous

inclusive, and especially throughout the whole of the Mesozoic epoch, no higher vertebrates are at present known with certainty than Reptiles, Birds and Marsupials: the first so far predominating in numbers, and as regards many species in size, as to justify the expression 'Reptilian Age' as applied to that period.

The evidence of Bird life mainly consists of footprints, though some of these—as on the Connecticut sandstones (if they be true *Ornithicnites* and not reptilian) point to creatures of gigantic dimensions. Others, *e.g.* the *Archæopteryx*, and the *Ichthyornis* with fish-like biconcave vertebræ, assumed most remarkable forms, so that possibly the secondary or Mesozoic epoch might have been characterized as much by Birds as Reptiles; while the abundance of the *Pterosauria*, from the size of a pigeon, or even less, to the extent of upwards of twenty feet across the expanded wings, shews an enormous development of flying Reptiles which may possibly have 'represented' a considerable portion of Bird life such as now obtains everywhere.

Lastly, on arriving at the Tertiary beds, so many forms of all kinds of Mammals—successive groups replacing one another chronologically in the various strata—have been revealed by the researches of geologists as quite to justify the analogous phrase, 'Age of Mammals,' as descriptive of the Tertiary period.

The evidence then, as far as it is derived from geological discovery of the development of Vertebrate life, is that each great group or type became the prominent feature successively in an ascending scale of organization.

If now we admit that each great type did appear thus successively and assumed such numerical proportions as to render it a conspicuous feature of the life of the period, and that it subsequently sank into comparative insignificance; then we must examine each group by itself and see how far Evolution is supported by details as well as by the period of incoming of the group in its entirety.

In doing this we must apply those principles enumerated above to the different groups as far as each will apply: and if we find Palæontology furnishing sufficient illustrations, we shall be proportionally justified in regarding Evolution to have been the process of Nature.

If we commence with Fishes, as being not only the lowest type of Vertebrates, but the earliest known of that sub-kingdom, the first observation will probably be, that fishes are as numerous now if not more so than ever before. But are existing fishes as a whole of a more advanced or of a more elevated character than those, say, of the Devonian period? That large and prominent groups of fishes have almost disappeared is obvious, such as the Cestracionidæ, the Chimæridæ and Ganoids generally; while the Teleostei, their modern successors, far outnumber the few lingering representatives of those groups. Indeed, the question might almost be put thus; Are the Teleostei of a higher grade than the Ganoidei and Elasmobranchii?

Here, then, we are brought at the very threshold of our enquiry to discuss the exact meaning of 'high' and 'low' grade, or what *is* an organ of more or less specialized character.

As in all other questions of this sort, a clear distinction must be held in the mind between what is *absolute* and what is *relative*: and it must be remembered that the majority if not all terms in biological science are relative only. Now the word *grade* is purely relative and applies to the different stages of development exhibited by the various members of a Type or Group: while an organ of but little specialization is one which performs at least two if not more functions, which are executed by separate organs in more highly developed forms.

Such terms as *modification* and *differentiation* will be found useful; and we might qualify the former as *horizontal* and *vertical*. Modification being the more general expression implies change of form through descent, with or without any change of function: so that 'horizontal modification' would imply comparatively slight change of form with little or no change of function. Vertical modification, that is *differ-*

entiation, implies a more considerable change of form, for the most part coupled with difference of function.

As an illustration of the former, we may take the Foraminifera. In them we have the two essential elements of sarcode and shell. The forms of the shell are innumerable, but no other differentiated elements are met with possessing functions other than those peculiar to sarcode and alike for all.

As an illustration of the latter, a simple instance of modification of structure with differentiation of function may be mentioned. The internal sac of the Actinozoa acts as stomach and lungs, while in the next group the Molluscoida the organs have become differentiated, so that a distinct alimentary canal is present and a decided though humble nervous system appears. When, therefore, such distinctions can be clearly made out, we can judge how far modification may or may not support evolution; for it may be here noticed that modification without differentiation, or what is here called horizontal modification, supports the idea of Retention of Types, while the other, *i.e.* true differentiation, supports the doctrine of Evolution. Both of these facts must be accepted by the Biologist.

Fishes.

In applying these remarks to the old and the existing Fishes, one feature stands out especially prominent as marking a conspicuous difference, and that is the 'heterocercal' tail being so prevalent in the palæozoic fishes while the 'homocercal' tail prevails now.

Of this organ we may note three prominent types. First the notochordal condition as in *Undina* and *Macropoma* of the Devonian, *Caturus* of the Solenhofen beds, and *Lepidosiren* which still exists; secondly the *true* heterocercal form which obtains in Megalichthys and the Sturgeon; and thirdly the tails of the existing Teleosteans, *i.e.* those usually called 'homocercal,' of which three varieties are seen, viz. the bifurcating tail of the Salmon, the rounded form of the Wrasse, and the pointed tail of the Conger. Now it must be carefully observed that these last are really 'heterocercal' anatomically, though in consequence of the development of the wedge-shaped hypural bones, the external form of the tail appears equal-lobed[1].

The tail called 'diphycercal' by Huxley appears more simple than any of the above, and represents the archetypal or ideal condition in being merely an axis tapering to a point around which is the fringe-like tail.

My object in discussing, perhaps at too great a length, the peculiarities of the tails of fishes, is to shew how greatly that important organ has been altered in the different groups. Indeed the entire skeleton of modern fishes might have been equally discussed at length. The general result, however, I would wish the reader to perceive is that modern fishes are far more truly 'Fish-like' than their palæozoic ancestors, and that this

---

[1] *Zeitschrift für Wissenschaftliche Zoologie*, Band 14, 1864, p. 81. *Ueber den Bau der Schwanzwirbelsäule der Salmoniden*, &c. Von Theophil Lotz.

change has been brought about (as illustrated by the example of the tail) by the slow process of Evolution.

Again we must also particularly notice the frequent occurrence of the notochord amongst the Crossopterygian and other palæozoic fishes. This is a very decided embryonic feature, and obtains in the Lepidosiren; though *all* the Teleosteans have well-ossified vertebræ, and so far are of a higher grade.

Now this feature of the persistence in the adult stage of what is embryonic only in more advanced forms of life of the same type of Vertebrate, goes a long way towards supporting the general doctrine of Evolution.

The argument derived from the study of the stages through which an individual embryo passes will be spoken of hereafter; but it may be now generally stated that whenever embryonic types are found at all, especially when preceding more highly-developed forms of the same kind or family; then, we have strong evidence in favour of Evolution.

In discussing the respective merits of modern fishes and ancient ones, it is interesting to find early representatives of the Teleostei living contemporaneously with the predominating Ganoids when in their prime; just as we have a few Ganoids still surviving at the present day, such as *Polypterus, Amia, Lepidosteus* and *Ceratodus*: for Prof. Huxley has shewn that the genera *Coccosteus* and *Pterichthys*, especially the former, seem to have been representatives of the Siluroid Teleosteans: though the persistent notochord points to an embryonic state or to a lower grade than that of the modern Teleosteans.

It must be noted, too, that although a maximum degree of development is not always coexistent with numerical importance nor associated with the greatest bulkiness, yet it is quite *en rapport* with the doctrine of Evolution to find the most highly differentiated fish to have appeared so early as the Devonian epoch, while many of them were of very large dimensions; and the fishes at the present day to be undoubtedly modified descendants, but assuming characters which place them in a different if not lower grade than the prevailing forms of the Devonian epoch.

Lastly, a comparison seems capable of being drawn between the lowest of the vertebrates and the lowest of the invertebrates; that just as the Foraminifera have undergone an immense amount of modification without much if any differentiation into higher grades; so, fishes have undergone a like large amount of horizontal modification without much vertical differentiation.

A like comparison may be also made with the Brachiopoda: for if the species of Terebratula, Rhynconella or Lingula be compared, *viz.* the earliest forms of each with the last existing; it will be found that a wonderful uniformity prevails, though comparatively slight variations exist, which give great diversity of modification, but apparently little differentiation[1].

Amphibians.

Let us now turn to the Amphibia. The highest group, Batrachia, has never been discovered

[1] See *Life, its Origin and Succession,* by Prof. J. Phillips, p. 93.

in the so-called 'Amphibian age:' but this group was mainly represented by gigantic Labyrinthodonts.

Now these conveyed a stronger savour of old piscine than of reptilian characters. Moreover they were associated with, if not preceded by, forms representing the other existing order, Gymnophiona; the history of which is unknown.

The Archegosaurus, however, with its persistent notochord, and the absence of the ossified occipital condyles which characterize the skull of higher forms of the Amphibia, clearly points to conditions embryonic, or at least of a low grade; while the details of the structure of the teeth, being as it were the rudimentary conditions of the Labyrinthodont type, together with the cephalic plates of the exo-skeleton, point to the old Devonian dendrodont fishes.

The Amphibia resemble fishes in the entire absence of an amnion, and in possessing for at least some period of their existence branchial filaments. Moreover they are somewhat sharply distinguished from at least birds and reptiles by the fact that the skull articulates with the spinal column by two condyles. On the other hand, those of the Amphibia which possess limbs have the parts of their limbs corresponding with those found in the higher vertebrates; whereas this is the case with no fish.

Prof. Owen has maintained that the affinity to fishes was more pronounced in the Amphibia than in reptiles, though that palæontologist recognises the group as intermediate if not a link between fishes and reptiles.

Prof. Owen thus speaks of them (*Palæontology*, p. 217): "The conformity of pattern in the dermal, semidermal, or neurodermal bones of the outwardly well-ossified skull of *Polypterus, Lepidosteus, Sturio,* and other Ganoid fishes with well-developed lung-like air-bladders, and in the same skull-bones of *Archegosaurus* and in the Labyrinthodonts; the persistence of the notochord and branchial arches in *Archegosaurus* as in *Lepidosiren*, the absence of the occipital condyle or condyles in *Archegosaurus* as in *Lepidosiren*, the presence of labyrinthic teeth in *Arche-*

gosaurus as in *Lepidosteus* and *Labyrinthodon*, the large median and lateral throat-plates in *Archegosaurus* as in *Megalichthys*, and in modern *Arapaima* and *Lepidosteus* :—all these characters point to a great natural group or series, shewing the gradations of development which link and blend together fishes and reptiles within the limits of such group. The Salamandroid (or so-called 'Sauroid') Ganoids—*Lepidosteus* and *Polypterus*—are the most ichthyoid, the true Labyrinthodonts are the most sauroid of the group. The *Lepidosiren* and *Archegosaurus* are intermediate gradations, one having more of the piscine, the other of the reptilian characters.

"The *Archegosaurus* conducts the march of development from the Ganoid fishes to the Labyrinthodont type, the *Lepidosiren* to the Perennibranchiate type.

"Both illustrate the artificiality of the supposed class-distinctions between fishes and reptiles, and the unity of the cold-blooded vertebrates as a natural group. There is nothing in the known structure of the so-named *Archegosaurus* or *Mas-*

todonsaurus that truly indicates a belonging to the saurian or crocodilian order of reptiles. The exterior ossifications of the skull and the canine-shaped labyrinthic teeth are both examples of the salamandroid modification of the ganoid type of fishes. The Ganocephala and Labyrinthodonta characterize the transitional period between the palæo- and mesozoic epochs."

Enough has now been said to shew how closely the Amphibia are allied to Fishes: and that the retention of piscine characters so frequently seen in the fossil Amphibia is explicable on no other theory than that of their gradual evolution from Fishes.

Reptiles.

The next group to be considered is the Reptilia.

The first known types (Permian Thecodonts and Lacertians) are of tolerably high grade, necessarily point to a lost ancestry, and might be taken as witnessing to the untenableness of Evolution: but, be it remembered, such a view would rest solely on negative evidence, whereas Evolution is based on positive evidence[1]. A group of far greater importance for our purpose is the Enaliosauria (Owen), of which Ichthyosaurus especially exhibits—as, indeed, its name reminds us—a union of characters derived from fishes on the one side and reptiles on the other.

Prof. Owen has remarked: "With the retention of characters which indicate an affinity to the higher Ganoidei, the exclusively marine Reptiles more directly exemplify the Ichthyic type in the proportions of the premaxillary and maxillary bones; in the shortness and great

---

[1] It has been accepted as a reasonable hypothesis that the line of passage from fishes to reptiles does *not* lie along the Amphibia at all, but through some group of which perhaps the Lepidosiren is the sole existing representative: and that the Amphibia are an offshoot terminating abruptly with the order *Batrachia*. Although having a common origin with Reptilia in the Fishes, yet the Amphibia as sharply differentiate from them in certain particulars, *e.g.* the fact that the skull articulates with the spinal column by two condyles, while the basi-occipital remains unossified.

If this be true the early ancestry of Reptiles might have been developing for ages during the carboniferous and earlier epochs; but either no remains have been left, due perhaps to there being neither exo- nor endo-skeleton; or else geologists have not met with them.

number of the biconcave vertebræ; in the length of the pleurapophyses of the vertebræ near the head; in the large proportional size of the eyeball with its well-ossified sclerotic coat; and especially in the structure of the pectoral and ventral fins... Thus the Ichthyosaurus presents a general type of structure more conformable with that of which the Archegosaurus and Labyrinthodonts manifest the phases of development, and in which the ascent from the gano-salamandroid fishes reaches its culminating point in Ichthyosaurus" (*Palæontology*, p. 220).

Buckland in his *Bridgewater Treatise* shewed how Ichthyosaurus possessed a generalized structure, for it had "the snout of a porpoise, teeth of a crocodile, head of a lizard, the biconcave vertebræ of a fish, the sternum of Ornithorhyncus, and paddles of a whale!"

As points of relatively low grade may be mentioned the unankylosed processes of the centra; the numerous bones of the rami of the jaws; while the teeth were arranged in long furrows instead of distinct sockets. The cranium itself, though resembling that of a dolphin, had the bones by no means similarly integrated.

We have seen that while fishes and Amphibia pass insensibly as it were into each other, yet the change either from fishes or Amphibia to reptiles is somewhat abrupt. Yet as a whole the Amphibia stand intermediate between fishes and reptiles. Very probably, however, they may not represent the actual line of ascent to reptiles, but are an offshoot by themselves; yet when we examine the Enaliosauria we at once recognise the piscine features, and as we know they had no exo-skeleton and that the Amphibia and older fishes often retained the notochord as in the Lepidosiren of the present day, so the forerunners of true reptiles may have had those two features combined, would therefore have left no remains at all which would escape destruction, and so the hiatus would occur which, as a point of fact, does exist.

Taking, however, Ichthyosaurus as we find it, there are presented to our observation certain affinities, which cannot be questioned, to fishes, and these have to be accounted for. Now Evolution is suggested as more probable than the Creative hypothesis, and that these affinities are undoubtedly due to descent with modification.

Birds.

It is an admitted fact by all comparative anatomists, that no sharp line can be drawn between Reptiles and Birds; while Prof. Huxley's discoveries in the osteology of the Dinosauria point with cogent force to the probable line of passage from one to the other[1].

The *Archæopteryx* and *Ichthyornis*, the sole fossil representatives of mesozoic bird life which afford any details for generalization:—a few bones in the Greensand strata, &c., and footprints, as in Connecticut Sandstone, being all the rest:—point to reptilian and piscine affinities, though they do not belong to the Ratitæ, to which the *Dinosauria* are affined, for the unankylosed condition of the metacarpals of the

---

[1] *Pop. Sc. Review*, Vol. VII. p. 237; and *Quarterly Journal of Geol. Soc.* Vol. XXVI. p. 12.

former and the great length of its tail are two characters which have considerable significance.

The general result of the evidence derived from the fossil remains, coupled with the points of affinity between living birds and reptiles, shew that there is an intimate connection between the two classes, and that no sharp line of demarcation can be drawn between them. Hence, if the evidence *per se* may be thought at all weak from want of abundance of material, as far as it goes it is all in favour of Evolution; and the very fact that the only known instances of birds having been preserved in the mesozoic strata should show such remarkable affinities is, at least, very significant and possesses a degree of importance quite unexpected. As it is, it furnishes *positive* evidence for Evolution; had it been otherwise, it would have been *negative* evidence only for the Creative hypothesis.

Mammals.

We have now arrived at the Mammalia, and of this group we do not meet with many specimens until we enter upon the Tertiary epoch, when they suddenly burst into view in a profuse and rapid succession of forms.

That all the Mesozoic mammals hitherto discovered were probably marsupial has considerable significance; and when coupled with the fact that wherever Mammalia have been found, even so widely apart as England, Germany, and N. America, they have always been marsupial, *i.e.* were *homotaxial* in mesozoic periods, the significance rises to a considerable degree of importance.

It may be noted here, that a similar degree attends the like discovery of Amphibia in the Carboniferous to Triassic ages, in the same countries as those in which marsupial life seems to have been homotaxial. On the other hand, an absence of all other kinds of Mammals in the first case, and of many Reptiles in the other, must not be forgotten.

It could not perhaps be maintained that such discoveries point to Evolution further than that they hint at the incoming of the higher forms of life respectively, which should culminate and take their position at the head of creation when the Amphibian and Reptilian ages should have passed away.

As soon, however, as tertiary strata are investigated, whether in this continent or in America, such an abundance of Mammals is at once revealed as to give a more certain note as to the probability of Evolution being true, in their case at least, rather than the old Creative hypothesis.

It will not be out of place to pass in general review some few of the principal types of mammalian life that prevailed during the successive tertiary periods.

In the earliest periods of the Tertiary epoch, very generalized forms have been discovered in both continents; such forms are Lophiodon, Coryphodon and Hyracotherium; in the first two of which are, as it were, blended the well-differentiated structures of Artiodactyla and Perissodactyla, the 'even toed' and the 'odd toed' groups, so thoroughly distinct now.

The Palæotheridæ of Eocene and Lower Miocene times well illustrate intercalary forms in an apparently strict line of descent, for they are succeeded by the Paloplotherium, Anchitherium (in America), Hipparion (Europe), and Equus, *forming a true successional series by differentiated modifications by descent, which renders the truth of Evolution a moral conviction.*

Again, the Hyracotherium of Lower Eocene appears to have given rise, or at least shews affinity, to the *Chæropotamus* of the Upper Eocene. This in turn finds modification in the *Anthracotherium* and *Palæochærus* of the Lower Miocene, and in turn passes upwards into Dicotyles of to-day, which is akin to another descent, Pigs, from the same source.

A similar result is obtained if we trace the apparent line of descent of the Rhinoceridæ; so numerous have been the forms throughout the Tertiary epoch.

Similarly, with regard to the Proboscidia, while there are but two species of existing Elephant, there have been discovered at least forty species of Mastodon and Elephas together, including the intercalary forms linking these genera.

Another remarkable illustration may be taken from the Carnivora. The form *Ictitherium* of the Upper Miocene of Pikermi had characters which combined the two now well-recognised sub-orders, Viverridæ and Hyænidæ, represented by the living Civets and Hyænas[1].

It is ascertained that about four-fifths of the fossil Eocene fauna belonged to the Perissodactyla, while one-fifth only is to be referred to the Artiodactyla. On the other hand, while the latter group is very abundant now, there are only three genera, excluding Hyrax, which still survive, *viz.* Rhinoceros, Tapirus, and Equus.

In thus reviewing the successive groups of Mammals, the most remarkable feature that the Palæontologist cannot fail to notice is the presence, so frequent is it, of intercalary or intermediate forms.

---

[1] Absolutely trustworthy tables of descent cannot be made, but the above (taken from Mr Gaudry's work) shew strongly in favour of 'descent with modification.'

Now it is again and again urged by the opponents of the doctrine of Evolution, that they are absolutely wanting in the geological records. This is simply a misstatement of facts. Opponents seem always to be alluding to *minute variations* from a typical species; referring tacitly perhaps to Darwin's theory of Natural Selection. But as far as the doctrine of Evolution is concerned, it is not that we look for intermediate forms in order to support the doctrine; it is that, finding such intercalary structures, the Evolutionist believes them to be more probably due to a community of descent than to any other cause. If minute gradations are to be found— and they do occasionally if not often occur— then they strengthen the reason for believing such to have been their origin, though are not absolutely necessary; for I do not pronounce as to the methods Nature has adopted in producing forms of life, but simply endeavour to bring forward facts, and to draw the most rational conclusion from them; that being expressed by the word Evolution.

Indeed, the existence of intercalary forms, or of beings partaking of characters belonging to two or more different animals, is a feature perfectly well known both to the Naturalist and the Palæontologist, including under the first, students of existing forms of life. We have already seen that Fishes and Amphibia are linked together, and although there is a sharper line in some respects between Amphibia and Reptiles, yet they are decidedly akin to one another. Reptiles, again, cannot be separated from Birds; while the Ornithorhyncus seems to be the last remnant of the line of ascent between Reptiles and Mammals.

In the next place, if we consider individual genera, the following will illustrate intercalary or combining forms as having existed previously to the Tertiary epoch:—*Archegosaurus, Ichthyosaurus*, the *Ornithoskelidæ* and *Compsognatha;* but in the tertiary strata they become, as we have seen, far more numerous: for example, *Lophiodon, Zeuglodon, Oreodon, Ictitherium*, and *Mesopithecus;* while *Metarctos, Amphicyon* and

Hemicyon, and others combined Dogs and Bears. *Hipparion* was an obvious link between the *Palæotherium* and *Equus*, and *Helladotherium* united Giraffes with Antelopes. Finally *Zeuglodon* linked Cetaceans with Seals; and *Oreodon* Ruminants with Pigs.

In North America, the discovery of numerous forms of mammalian life has shewn features identical in character, though even more strongly pronounced than is the case with those of Europe, *viz.* in being for the most part less specialized in the earlier strata than in the subsequent ones. As a most remarkable illustration of a highly generalized form, and which at the same time is intercalary between Pigs and Ruminants, is the Oreodon. It possessed the molar teeth of ruminants, the ulna and radius of a hog, the cranial and temporal regions of a camel, and in other respects resembled the *Cervidæ*.

Similarly, the Elotherium linked the existing hog-peccary and hippopotamus.

Such, then, is somewhat of the evidence in

favour of Evolution, derived from the study of the mammalian forms of the Tertiary epoch. In other words, fossils exhibit certain remarkable features which are represented by the expressions, 'more or less generalized or specialized structure,' 'average forms,' 'transitional or intercalary features,' &c.; and the generally received idea is that this has come about by community of descent with modification, or, what is termed, differentiation—the general process being called Evolution.

It may be observed that whenever considerable numbers of specimens of a particular type or group have been found, there are sure to be many forms, which can be selected as exhibiting such transitional features between two or more well-marked forms, and linking together either contemporaneous or successive beings.

Now it cannot be denied that these intercalary forms are best accounted for by the theory of Modification by Descent, or Evolution: for though, as Prof. Huxley remarks, it is impossible to say which of three forms gave rise to the others, or whether all three may not have risen from a fourth which preceded them; yet the fact of each form not being, as it were, a decided entity of itself and well distinct from any other—frequently there is a closeness of agreement between them and not unfrequently are they connected by minutely transitional forms—shews strongly in favour of Evolution or Community of Descent; and this is all that the doctrine contends for.

It is perhaps desirable to note a difficulty that might be felt in the discovery of a highly generalized form in a later period than that in which the specialized structures, combined in it, had already become differentiated in separate individuals.

The explanation is simple enough. It is this:—that such generalized forms continued to transmit to their descendants their own characters without any or with but slight modifications; while the particular causes which produced differentiation elsewhere were never brought to bear upon their individual posterity.

Moreover, the more generalized the form the more presumably probable is it that it may continue to live with others differentiated from it: for as a rule generalized forms would be more adaptable to a variety of conditions and consequently less liable to extermination.

It may be desirable to conclude this portion with the following observation.

In speaking of types appearing successively in order of development, it must be understood that such order refers only to types which can be presumably arranged in a lineal series: while two or more parallel series might have existed cotemporaneously which had the same origin or parentage. Thus, the Pterosauria and Dinosauria are two geologically cotemporaneous groups or series of Reptiles, the former terminating with itself; the other possibly becoming the line of passage into Birds, which in turn has terminated that series: a third line, to which perhaps the Ornithorhyncus may point, has probably given rise to Mammals.

It would seem not out of place to state here

what may have been the origin of these as well as Invertebrate *Types*, for we have seen that Geology lends strong support to the idea of each type having come in successively and having taken as it were a place at the head of creation. So that, as types are only strongly pronounced grades or differentiations, it may be perhaps correct to say that *want of competition*, or a more perfect freedom from a struggle for existence, may have given rise to types. If this be true, it will apply amongst other things to the isolation of the genus *Homo*.

A chapter on Man will be found further on.

CHAPTER V.

INVERTEBRATE TYPES.

If the fact be considered established that the great groups of Vertebrate life succeeded each other so as to become the leading representatives of the animal creation successively in an ascending scale of organisation, then, on the same principles, we may reasonably expect to find the highest group of the Invertebrates to have arrived at its culminating era in the lowest Palæozoic ages or even long before them.

Now we may first notice that for invertebrate development we have an enormous period at our disposal; for the thickness of the Cambrian and Laurentian strata is probably greater than all the strata overlying them put together.

Now as a matter of fact we do find the Mollusca at the Silurian epoch to have arrived at a maximum of development in their Cephalopoda, numerically and in bulk if not in differentiation of structure[1]. Here, then, one supposed stronghold of the opponents to the theory of Development—that, according to them, no such high and perfect types should have existed at that age—is on the contrary *one of the strongest proofs of Evolution.* It shews that probably the huge Cephalopods were conjointly with crustaceans (*e.g. Euripteridæ*) the sole predacious monsters of the deep of that age. At all events we find these types excessively abundant and characteristic of those times, and, according to the views of Evolution advanced in these pages, it is just what one would have expected[2].

[1] It has lately been shewn that the many forms of straight and uncoiled *nautilidæ* represent more or less undifferentiated conditions of young *nautili*.

[2] Although opponents of the theory of Evolution so often point to palæozoic ages as affording no evidence in support of it, yet since large suites of Trilobites, and numerous forms of Crustaceans have been found, some significant tables of descent have already been attempted.

The Laurentian strata are too barren to reveal the progress of the other kinds of Molluscs: but it may be borne in mind that as the highest group, Cephalopoda, were expected to be and actually were the main feature of the early palæozoic, we should expect to find all other and lower forms of Molluscs—not to add all other still lower forms of life—differentiated long before, and their representatives living cotemporaneously with the Cephalopods, as in fact is exactly the case.

On the other hand, comparing palæozoic invertebrates with mesozoic and recent forms, it is a well-recognized fact that they are all, without exception, of inferior grades.

Descending the scale of life we find that the two laws of maximum of bulk and lowness of grade will at least hold good with the earliest known fossil, *viz.* in the gigantic Rhizopod, *Eozoon Canadense.*

Of many groups of Invertebrata few or no records are to be found, and consequently one can argue nothing directly, such as of insects,

&c. And therefore, because they and other large groups may now furnish well-differentiated sub-groups, genera and species, there is nothing to warrant the statement that such existing types nullify Evolution. Such a supposition can only be based on the negative assumption that intermediate or intercalary forms never existed. But the frequent discovery of such in other groups better capable of being preserved, as the Molluscs and the Tertiary Mammals, tends to support the belief of the probability, if not moral certainty, of their having once existed.

With regard to intercalary forms among at least certain of the Invertebrata, the evidence exists probably in more instances than amongst the Vertebrata: for whenever large suites of fossils, of forms approximating each other, can be procured, minute gradations will very generally be found.

It is only when a few only have been discovered that differences are conspicuous, and so genera and species are made: but as soon as large numbers are procured, the supposed species often vanish as distinct entities, and not unfrequently the genera disappear too. Thus, of species of Molluscs, *Mya arenaria* and *Mya truncata* blend together through intercalary and intermediate contemporary forms. And as an example of genera differentiated from a common stock may be mentioned the two well-marked forms, *Purpura lapillus* and *Fusus antiquus*, so distinct at the present day; yet they appear to blend together by intercalary forms in the Red Crag deposit[1].

As a matter of fact all conchologists appreciate the same difficulties of classification as are felt by the botanist. Like genera amongst plants, whenever large suites of shells are procurable whether living or fossil, species apparently very distinct when seen apart, are frequently found to be united by a host of intercalary forms.

For example, any one critically examining the immense number of *Terebratulæ* of the Oolites,

[1] I am indebted to Mr E. Charleworth for calling my attention to this fact, which I have since tested.

the *Fuci* and *Volutæ* of the Eocene, *Tellinæ* of the Crags, &c., will soon discover how hopeless a task it is in many cases to distinguish the species satisfactorily.

That several groups of the Invertebrata have remained from the earliest ages without much or even any differentiation is in no way adverse to the theory of development. Thus of Corals, they have changed to some extent in form; whole groups having disappeared, such as the *Rugosa*, while others have taken their place; but the modern corals do not exhibit much differentiation or advance in structure over the palæozoic forms. The same remark applies to the Foraminifera and sponges.

This feature of retention of type is to be found in many groups of life, including plants: thus amongst the carboniferous flora the ferns agree in all essential features with modern ones, even if some genera (*e.g. Pecopteris*) be not still living! Similarly the fructification of *Calamites* closely agrees with that of *Equisetum*, while Mr Carruthers has shewn how exact is the similarity between *Flemingites* and *Lycopodium* as well as that between *Triplosporites* and *Selaginella* (*Sc. Opinion*, 1869, p. 150).

Retention of type is a principle as wide in its bearings as Evolution, and must be fully recognized and appreciated by any exponent of the history of life. For, while Evolution or modification by descent accounts for all the variety of forms of animals and vegetables, yet persistence or retention of type, with or without 'horizontal modification,' accounts for the fact of lower grades and types coexisting with higher and more perfectly differentiated forms.

These two principles of Nature must be accepted conjointly, otherwise only a partial view of the phenomena of Nature will be obtained, and the real truth as to Nature's methods will be greatly obscured.

CHAPTER VI.

Negative Evidence.

If we now regard the successive appearance of vertebrate types as in any way corroborative of the theory of Evolution, we must bear in mind that the earliest fish, amphibian, or reptile, that may have been discovered, is not always of low organization relatively to the group of which it is a member. And, if Evolution be true, those individuals, such as the Thecodonts of the Permian or the Onchus of the Silurian, must be tacit witnesses to a line of ancestors of unknown length. In other words, we have no grounds for presuming each successive type or group to have come into existence *abruptly*, but rather that it overlapped as it were the culminating period of the group next below it in type.

It is worth while, therefore, to repeat the necessary caution not to overrate the value of negative evidence. All that can be contended for is, that geological discoveries hitherto lead us to suppose that each great group of vertebrate types became in turn the characteristic feature of animal life; and they were so chiefly in the two phases of *numerical preponderance* and of *bulk:* not however necessarily being representatives of relatively high grade in structure, though this may sometimes have been the case. Thus sharks appear to have been a prominent piscine type in former days, and the allies of the sturgeon are not of low development, while at the present day the Teleosteans which furnish the predominating types, if not somewhat degraded, are not much differentiated. On the other hand, there is no evidence as yet of the true Batrachia, or highest form of Amphibia, until the Triassic period is long passed.

The fact that at least some one or more members of each group should culminate in size or bulk at some period of its evolution , appears to be a very significant fact; for a justifiable inference may be drawn from it, namely, the probable absence of any intense struggle for existence with other animals of equal size and strength, while the negative evidence of such animals never having been found supports the idea. Again, homotaxy points in the same direction; for although it might be urged that bulky animals of one kind might have lived at one spot, but smaller ones elsewhere might have been associated with large ones of a different kind, yet it must be remembered that while large Amphibia lived in England, others as large lived in America. The same is true of the mesozoic reptiles, including gigantic Pterosauria.

With reference to the Mesozoic reptiles we may remember that the lowest forms of Mammalia —the Marsupialia and Edentata—did not attain their maximum of development, as far as bulk is concerned, until probably a very late tertiary period; thus indirectly leaving us to assume that there were no other mammals of equal size to compete with those *reptiles.*

The bulkiness of an animal *per se* depends probably upon abundance of food and ease in procuring it, yet so true is the law of development in growth to a maximum, and then of decrease to smaller dimensions, that it seems dependent upon some far-extending principle which affects all creation, plants as well as animals.

A few illustrations will recall this fact to mind. The Eozoon is a gigantic rhizopod. In the early palæozoic ages were gigantic Cephalopods and Crustaceans. Enormous Fishes then succeeded, to be followed, in the ages which linked the primary to the secondary epoch, by huge Amphibia.

Then again huge Reptiles, marine, terrestrial, and aërial, followed, only to disappear and yield their place to Mammals, many of which vastly surpassed in bulk their representatives of to-day, such as the Megatherium as compared with sloths, the Dinornis amongst Birds, and the large Marsupials of Australia, and lastly, if Genesis reports truly, man himself was subject to the same law; for we are distinctly told "there were giants in those days," an expression evidently implying much more than the merely casual occurrence of them as is often the case now.

Similarly has this law governed at least certain of the groups of plants, for the Lycopodiaceæ and Equisetaceæ of the present day cannot boast of members of anything beyond humble herbs; yet we know of what importance their gigantic prototypes were in the carboniferous era.

This law, therefore, of attainment to the maximum degree of bulk points in part to a want of a struggle for existence, *i.e.* to an absence of other forms of life of equal size, and therefore is an indirect witness to the statement that at the period of its greatest abundance each group was, so to say, at the head of creation: and that as the next type—already begun long before the preceding had waned or perhaps begun to wane—developed, multiplied, and, following the same law, gradually replaced the preceding, and took its turn as supreme.

There is yet another fact worthy of note which applies at least to the Reptilia, and which would seem to confirm the suspicion that few other animals of a higher grade existed contemporaneously with them.

Whenever a balance of life is to be maintained amongst a variety of mammalian forms at the present day or in the tertiaries, it has always been effected by the carnivora keeping down the herbivora; but where a single type prevails, certain members of that type assume the carnivorous forms and functions, *e.g.* the Dasyures amongst the Marsupials of Australia represent the carnivorous type; the Bandicoots, the insectivorous: the Phalangers simulate the Quadrumana, while the Wombat has the peculiarities of a Rodent; the Kangaroos, of the Ruminants; and *Antechinus minutissimus* affects the form and habits of a mouse!

So, conversely, when we find gigantic Dinosaurians of almost elephantine size, terrestrial and herbivorous, assuming the functions now performed by large mammalian quadrupeds, as well as many kinds of Pterosaurian reptiles, from a few inches across to upwards of twenty feet, supplementing if not superseding Bird-life, we may with some reason infer that this compensation of functions points to a like absence of mammalian and perhaps ornithic types.

CHAPTER VII.

HOMOTAXIAL LIFE.

It will be advisable to add a few more words on the principle of *Homotaxis;* for this reminds us that the evidence is not based upon geological discoveries in any one locality by itself, but that wherever geologically synchronous deposits have been examined in different countries or continents *homotaxial* life has ever been discovered. That is to say, though no longer can the notion of identity of fossils indicate necessarily synchronous strata, nor can synchronous strata be known by an identity of fossils, yet representative forms will be more probably found, *i.e.* some other genus, species or race of the same group.

This law of homotaxy appears to be of more

importance than is usually assigned to it as evidential of Evolution. For, whenever abundance of individuals of a group prevailed to a considerable extent, if not to the exclusion of others, we are more likely *ceteris paribus* to find evidence of it; and therefore we may reasonably assume such groups as have been mentioned to have been characteristic of their respective periods.

An objection has been raised that they were so of certain localities, but not necessarily of periods: thus, it cannot be denied, for example, that local circumstances may have determined the occurrence of certain prevailing types, *e.g.* the Amphibia frequenting the marshy districts in which the coal-plants grew. But when we remember that such particular examples as of the Irish coal-fields at Kilkenny were neither the only ones nor the most representative, but that the huge Labyrinthodonts which traversed the margins of the great salt-lakes or inland seas of the Triassic period stamp that transitional time as specially characterized by Amphibia;

and, moreover, when we find the law of homotaxy hold good for the same kind of animal in the American strata at the same periods—as is also true for all other kinds of life in their respective strata and periods—surely the inference is at least strong that the true reptiles were in a minority at that period when the Amphibia flourished, and not so characteristic as they; so that the Amphibia were not merely characteristic of place, but also of time.

In bringing forward this law of homotaxy a distinction must be drawn between *identical* and *representative* forms. At the present day and throughout the later geological epochs, *i.e.* the Tertiary at least, representative forms are more usually found in widely separated countries, though in the earlier and palæozoic times many forms appear to have had a cosmopolitan existence on opposite hemispheres. This law is more especially true of relatively lower grades of life: and as certain groups of earlier ages were of lower organization than those which succeeded them of the same type, *e.g.* the Crustaceans and Brachiopoda, so correlative with this fact is the greater tendency to diffusion without much modification.

William Smith, as is well known, first propounded the idea of identical fossils being found in distant but corresponding strata, or those of the same age. But, although this cannot be always maintained, yet homotaxial life will be found to be true, or the *facies* of a fauna in one part of the world closely resembles that of another found in synchronous strata. Thus, the Trilobites of England are paralleled by those of America, Sweden and Bohemia, and the fauna of the British carboniferous limestone finds its close representatives in that of South America, China and Australia.

It may be observed here, that speaking generally there is less diversity of form in those homotaxial faunæ of palæozoic ages than in later periods. In other words, the early ages of the world saw more generally diffused types; and this is correlated to the fact that the more diffused is a type so proportionally is the grade low of the individuals. This is true both of animals and plants. Of the latter the wide-spread carboniferous forms will suffice to illustrate the principle.

As we ascend the scale of life and approach more modern times, faunæ become more and more restricted, and more representative at distant places, but of similar climate and other physical conditions.

We may observe also how even physical (inorganic) conditions seem to correspond to these laws. Lithological characters are far more uniform and more widely distributed amongst the early formations than amongst the later. Thus the limestones of N. American Silurian rocks are undistinguishable from those of Shropshire, and the carboniferous limestone has an identical appearance in whatever part of the world it may be observed.

Moreover, the continuance of similar physical conditions appears to have been as prolonged as they were generally diffused in olden times; for the thickness of the earliest strata or formations is enormously greater than that of later times, which are characterized by much more varied and highly developed faunæ.

Thus the Lower Silurian strata alone constitute 25,000 feet, while the entire Cretaceous formation is little more than 1100; and the Laurentian probably equals all the subsequent strata taken collectively in thickness.

CHAPTER VIII.

SUMMARY OF THE EVIDENCE OF EVOLUTION FROM EXISTING BEINGS.

THE evidence advanced by Mr Darwin in his *Origin of Species* and other works, and by Mr Wallace in his contributions to the theory of Natural Selection, is of course just so much evidence for the general doctrine of Evolution *per se*, when we leave out of sight that theory altogether. Thus 'heredity with variation,' or a general likeness in the offspring to, but with individual differences from, their parents, is a fact which no one will now gainsay: notwithstanding that persistence of type may hold good in many cases with or without any *horizontal* modification at all.

Secondly, no one can pronounce to what

degree the variations in the offspring may amount, provided sufficient time be allowed in which many and successive generations may be born.

A very frequent assertion by opponents of the doctrine of Evolution is that species are definite and actual entities, capable of a certain range of variation, but never passing into other definite and actual species: though it is true they very generally mean contemporaneous species. Now the theory of Evolution never countenances the idea of any two species, of *the same relative generation* in descent, passing the one into the other; but if it be maintained that no true species can ever give rise by descent with modification to another true species in a subsequent generation, then the burden of proof lies with the opponent.

On the other hand, not only are specific forms, *i.e.* if seen alone, so far distinct as to be worthy of the name of species, frequently united by intercalary forms; but enough differentiation has been produced by artificial cultivation or breeding by selection to warrant the supposition that if these offspring had been found wild and their parentage unknown, a naturalist would have at once pronounced them not merely of distinct species but of distinct genera.

Hence the whole question as to the probability of Evolution of all organisms being true may be made to turn upon that one point, *viz.* Can the successive offspring differ ultimately so much both morphologically and physiologically as to constitute a new species or even genus, *i.e.* what would be, at least, reckoned such according to the most approved standard?

Mr Darwin distinctly tells us that in Pigeons such a result has actually been obtained so far as morphology is concerned, but as yet all of the same kind breed together; so that the physiological barrier of sterility has not yet been set up.

Again, the long-cultivated cereals with no known wild counterparts point to the same conclusion; as also do those genera of wild plants which can boast of large numbers of subspecies or varieties, the existence of which cannot be satisfactorily explained on any other hypothesis than that of community of descent with modification. Such genera may be named as *Cassia, Rosa, Rubus,* and *Salix.*

It is beside the main question, as has been already pointed out, of the doctrine of Evolution to enquire whether the different forms appear suddenly or by slight modifications. The latter condition is required by the theory of Natural Selection. But it is immaterial for the doctrine of Evolution itself, provided such intercalary forms can be shewn to exist or with some degree of probability to have existed.

That they do exist in many cases is a fact admitted by everyone who has attempted to classify a large series of specimens of allied species or genera. Dr Hooker and Mr Bentham[1] have expressed this opinion; so that classification, once thought so easy with the facile scheme of

---

[1] Speaking of the *Asteroideæ*, a tribe of the *Compositæ*, Mr Bentham says:—"nearly the whole of the 90 genera, comprising above 1400 species, pass into each other through intermediate forms." *Jl. of Lin. Soc.*, Vol. XIII. p. 402.

Linnæus, has now become a task which no mind of ordinary ability can undertake with much chance of success. Rare powers of discrimination, with due appreciation of relative values and great breadth of view, are essential qualities of the systematist. The study of groups in all their minute ramifications of affinity has ere now forced the doctrine of Evolution upon the attention and acceptance of the systematic botanist.

Similarly is it the case with the Conchologist. Mr Jeffreys alludes to the transitional characters of shells so often found linking two or more species if not genera together, while aberrant forms of the same species, frequently caused by difference of locality, are often so peculiar as to be called by different specific names.

We must also bear in mind that in classifying plants by their morphological or physiological affinities, we are occasionally completely out of our calculation, for the two phenomena though generally correlated are sometimes quite distinct. Thus the three supposed genera of Orchids,

Monocanthus viridis, *Myanthes barbatus*, and *Catasetum tridentatum*, are one and the same species, the first being male, the third female, and the second possibly neutral, having appeared on the same spike of flowers with the first and third respectively.

On the other hand the three genera *Rhodora*, *Azalea* and *Rhododendron* will interbreed, and therefore physiologically are one genus, though morphologically distinct.

Surely, then, any opponent of the doctrine of development of species, who asserts that species are absolutely distinct entities, may be called upon, not only to define his idea of a species, but to explain the above phenomena in accordance with his suppositions, as well as to explain the many capricious unions and 'refusals' offered by morphologically distinct and closely allied plants, respectively.

That the barrier of sterility has not been established amongst cultivated plants and animals does not appear in any way to nullify the general argument. For fertility of crosses has been clearly shewn to be due to constitutional rather than morphological agreement. Plants (such as species of *Hippeastrum*) of the same constitution interbreed freely, while aquatic and land plants of the same genus refuse to do so.

But domesticated animals and plants are subjected to very different conditions from those of their wild state; moreover the external conditions of all cultivated or domesticated beings are so very similar that those exclusive peculiarities of wild species never find place amongst them. Moreover, with regard to domestic animals, there is a totally different mental as well as bodily condition, from that of those in the feral state, for they are not subject to laws impelling them to procure their own food, and to be wary in escaping from enemies, nor are they subject to the fear of their enemies, while many animals will not breed in confinement at all, and so forth. Hence, as all these complicated conditions of maintaining their existence in the wild state are excluded from the domesticated, so we cannot say but that the law of sterility may become inoperative in the domesticated state.

On the other hand, when we study the inter-crossing of plants, such capricious results are observed that nothing beyond the very general expression of 'sterility being for the most part proportional to distance of affinity' can be maintained.

Dean Herbert observed of hybrids that their offspring vary from absolute sterility to a greater degree of fertility than is possessed by the parents themselves: sometimes widely (morphologically) distinct species cross, as in the *Amaryllidaceæ*, yet closely allied species refuse to do so.

PART II.

EVOLUTION AND RELIGION.

CHAPTER IX.

MAN.

HAVING devoted the first part of this essay to the establishment of the theory of Evolution of living things, it would seem desirable to treat of the latest and most remarkable being, namely Man; and, as he is the only *moral* being known, a chapter on Man is an appropriate commencement of the second part of this essay, in which I propose to treat of the bearings of Evolution upon Natural Religion and Christianity.

Before however offering any remarks upon mankind, I would wish to state distinctly that I do not at present see any evidence for believing in a gradual development of Man from the lower animals by ordinary natural laws; that is, without some special interference, or, if it be preferred, some exceptional conditions which have thereby separated him from all other creatures, and placed him decidedly in advance of them all.

On the other hand, it would be absurd to regard him as totally severed from them.

It is the great degree of difference I would insist upon, bodily, mental, and spiritual, which precludes the idea of his having been evolved by exactly the same processes, and with the same limitations as, for example, the horse from the palæotherium.

Man offers three features of contrast with other animals, his body, his mind, and his moral faculties.

With regard to Man's bodily structure, Professor Huxley has shewn conclusively (*Man's Place in Nature*) that in no organ whatever is there a greater disparity between it and the corresponding organ of some one or other ape, than is the disparity between a higher and lower grade of the *Simiadæ*; but Professor Huxley adds that the absolute difference in its totality between Man and Apes is very great indeed. These are his words:—"The structural differences between Man and even the highest Apes...are great and significant. Every bone of a Gorilla bears marks by which it might be distinguished from a corresponding bone of a Man; and in the present creation at any rate no intermediate link bridges over the gap between *Homo* and *Troglodytes.*"

The conclusion the Professor arrives at is, that although there is a "no less sharp, though somewhat narrower, line of distinction between the Gorilla and the Orang, or the Orang and the Gibbon...yet the structural differences between Man and the man-like Apes certainly justify our regarding him as constituting a family apart from them...but afford no justification for placing him in a distinct order."

Now it appears to me that an important observation may be made here, namely, that if an animal or plant exists as the sole representative of a family or other large group, the principles of Evolution induce us to regard it as the last lingering remnant of a previously existing and extremely ancient type[1]. Thus amongst animals we have geological evidence of this truth shewn in the history of the *Cestracionidæ*, now represented by the Port Jackson Shark, in the few *Chimæræ* that have survived, and in the *Nautilus*, the sole representative of the numerous Tetrabranchiata. Of genera of which we have no history, there are the *Ornithorhynchus* and *Echidna*, and, what is more to the point, the *Cheiromys* of the order Primates.

Now this species appears to be especially significant, because it does not represent a high form, but, just as one naturally expects, a more generalized organism of a lower grade. But Man too is a genus of the same order Primates,

[1] It may be noted that this principle applies with equal force to the Vegetal Kingdom. The fact that species of insular floras are few in proportion to genera, and genera few in proportion to orders, points to the gradual elimination of many, with the retention of only a few forms which constitute the sole representatives of a lost ancestry. Such plants as *Fitchia*, *Pringlia*, *Welwitschia* and others illustrate this, while *Equisetum*, *Lycopodium* and others are known to have been represented in early geological ages by large and various forms.

constituting a family by himself, yet so far from being a generalized type, or of low grade, he is of *exactly the opposite character, viz.* of the very highest grade, most specialized, and of the highest type[1]!

Here then the genus *Homo* in its bodily structure stands out as a direct violation of a broad principle of Evolution.

There is another feature with regard to Man considered solely as an animal, which seems worthy of consideration, namely, that whenever single genera represent a family, there is no great tendency to much, if any, differentiation, or migration; the isolated genera of animals are restricted to certain areas, and appear to have lost the power of specific evolution.

With Man it is exactly the reverse. Not only is he adapted to every climate—it is questionable whether such is the case with any other species of animals—but he has peopled every quarter of the globe; he is truly cosmopolitan, and

---

[1] It is true man's affinities are dispersed through the *Simiadæ*, but in his totality he is decidedly first.—See a paper by Mr St G. Mivart in *Pop. Sci. Review*, April, 1873.

in this feature more nearly resembles certain plants. On the other hand, he has produced many well-marked and decidedly differentiated races.

With regard to further distinctions between Man and animals, and especially to points which cannot apparently be accounted for on any hypothesis purposing to explain the action of Evolution, I must refer the reader to remarks by Mr Wallace in his interesting work *On Natural Selection*, p. 303 *seqq.*, only mentioning here those particular features which are most striking.

The first Mr Wallace alludes to is the size of the brain of Man, saying that the brain of a savage is larger than he needs it to be.

The second point Mr Wallace alludes to is the 'Range of intellectual power in Man.' This is enormous in a highly-trained mathematician or man of science when compared with that of a savage who can only count four; yet the capacity of the brains would be the same; thus clearly shewing that the brain of a savage is much greater than he requires.

Mr Wallace next compares the intellects of savages and animals, and shews that practically they scarcely exhibit any advance above many intelligent animals: yet they possess a brain which has a potential intellect capable of being developed far beyond actual requirement.

He next passes on to the use of the hairy covering of mammalia, and notices the constant absence of hair from certain parts of Man's body as a remarkable phenomenon, and one which a savage keenly feels. Hence he draws the conclusion that Natural Selection could not have produced such a result, because Natural Selection demands alterations to be useful and not injurious.

Other points to which he alludes as throwing great difficulty in the way of developing Man from the Quadrumana by Natural Selection, are the hands and feet. Of the hand he particularly remarks: "It has all the appearance of an organ prepared for the use of civilized Man, and one which was required to render civilization possible."

He then adds the following significant words: "If it be proved that some intelligent power has guided or determined the development of man, then we may see indications of that power in facts which, by themselves, would not serve to prove its existence."

Other phenomena equally striking, and the existence of which is quite as incomprehensible if supposed due to Natural Selection, are speech and the power of framing abstract ideas. But what transcends all other distinctions is the conception of a God. Here the line of demarcation becomes sharp and clear; while the Moral Sense and Conscience involving the idea of Duty make the separation complete.

My object in quoting the above passages from Mr Wallace's able and interesting book, is to shew that I am not single in believing Man to have required some additional impulse beyond what Natural Selection or any other process of Evolution can furnish. That while he bears about his body rudimentary organs apparently so convincing of Evolution; while his intellectual powers can be paralleled in kind amongst animals, though in morals he stands apart; yet, there is the enormous gap between him and them which all observation and all philosophical reasoning has failed to bridge over.

It is in consequence of this gap that I would argue that Man cannot have been evolved solely by Natural laws, at least such as we are acquainted with in the Evolution of plants and animals.

I purpose now adding some additional remarks on the intellect of Man to the few I have brought forward from Mr Wallace's book.

If Man had been slowly evolved from the Apes his intellect must have gradually developed in proportion as his physical powers diminished, as witnessed by the degradation of the canine teeth and temporal muscles: and by the time the genus *Homo* had become established we should expect to find evidences of his semi-intellectual work, crude no doubt, but none are forthcoming. This, however, of course does not preclude the possibility of their yet being discovered. What, however, is more important than such negative evidence is that we should expect to find in the intellectual development of the children of civilized Man traces of that semi-intellectual condition, whatever it was, that was possessed by the supposed pre-human beings.

Surely the dawnings of intellect in a child ought to give us glimpses of that intermediate stage of Man between *Troglodytes* and *Homo?* The closest scrutiny, however, gives us nothing at all illustrative of any transitional stage between the jabbering of an ape and the intelligent utterances, however feebly expressed, of an infant of two years old, or even of only eighteen months existence.

Again, any comparison will shew children in many respects to be remarkably like modern savages. Now, savages of all tribes belong to the same species of Man as the most intellectual of European or other advanced nations; nay more, as far back as the palæolithic ages, if we may draw conclusions from similarity of workmanship, the mental capacity of Man must have been much the same as that of any savages of to-day: they were then, as all modern races are, of the same species of *Homo*.

In comparing savages to the children of civilized communities, all that is meant by saying that they are on the same intellectual plane, is that the best sample of an untutored savage may be fairly compared with a modern European child. Thus the intellectual powers appear to have been *arrested* at the stage characteristic of a child, in that certain savages cannot count beyond a few figures. There is the same adoration of physical strength and animal courage as in the modern schoolboy who, like the savage, as a rule[1], but little appreciates the greater importance of moral force. Thus, for example, there is the like instant exhibition of pugnacity on real or fancied grounds of offence, an accidental injury by a playmate being immediately and angrily resented. Again,

---

[1] But just as the *sanctity* of Truthfulness is held by some few savages, so is it by some boys.

the love of finery in the male sex so prevalent among savages is a feature which was characteristic of an age antecedent to the present one, of our own country: and of a time when the intellectual development of the community was at a very low ebb.

As a broad rule I think it a safe assertion to maintain that the love of finery in the male sex is in inverse proportion to the extent of development of the intellectual powers.

A savage therefore is equivalent to what might be called a child-man, an adult in physical development, but with the higher powers arrested at the stage of boyhood.

He may be compared to the *Welwitschia* amongst plants, which is an embryo in physiognomy but adult in physical characters, or as a fish is comparable to an embryonic amphibian.

That there is the same potential intellectual energy in the lowest savage as in a child of civilized parents is proved by innumerable cases of ordinary savages so placed in childhood as to have all the advantages of civilized life. Any one who listened to the intelligent conversation of the twin-negress or 'two-headed Nightingale' will corroborate this.

Mr Darwin, too, mentions how he was struck with the general intelligence displayed by the Fuegian who was taken on board the 'Beagle': for the Fuegian may be regarded as one of the very lowest grades of savages.

The same arrest of development in intellectual power produces similar or analogous results, allowance being made for differences of environing conditions; in the ordinary navvy or uneducated rustic, there is the same tendency to pugnacity, as also amongst pitmen and the Irish, the same idea of physical prowess, the same boyishness or even childishness of behaviour to one another in little ways which an attentive observer may often see, all witnessing to the same phenomenon of retention or arrest of a low or childlike intellectual grade in the adult.

The conclusions, then, I contend for are the following: that there is the same amount of potential intellectual energy in the lowest savage and in the civilized parent's child, and which only requires the occasion to develop it; that there is nothing in the development of the intellect of a child of civilized parents to point back to a supposed intermediate stage, had Man been slowly developed; for before a child can speak, its utterances are unmeaning, which is not the case with that of animals; for every cry of an adult animal is probably significant of some purpose. As soon as a child can speak, say from eighteen months, it is the dawning of *the intelligence of Homo*, not of a semi-homo. There is no intermediate stage between the infant's meaningless babblings and the thoroughly conscious and intelligent expression by which it denotes either parent.

I maintain that the powerful law of analogy ought to help us at this point, if it be true that the intellect of Man was slowly and 'naturally' developed from that of Apes.

It, however, here breaks down entirely and most significantly.

Again, in a comparison between Man and animals, it must be clearly understood that a very considerable amount of reasoning power resides in the animal world, though apparently somewhat capriciously distributed: yet the gap between all such displays of intelligence as one reads of in ants, birds, dogs, horses, &c., and that of Man is so enormous, that I maintain there are no facts in nature to corroborate, or even furnish, the idea of Man's intellect having been slowly developed.

'Nature makes no leaps,' is a sentence of profound truth; and if any leap be discoverable, it is to be seen from the intelligence of animals to that of Man.

To say that the lowest savage shews but little more intelligence than the highest Ape does not destroy the argument, for the gap is not so much to be measured by the actual display of intelligence as by the *Potential Intellectual energy* possessed by Man as a being, compared with any other species of the Primates.

The capacity of unlimited display of intellect is in the species *Homo:* though it may be dormant in many individuals. There is no such capacity in any animal known. All instances of reason circulate on a comparatively low plane *utterly incomparable with those of Man.*

Moreover, a question may be fairly raised as to the truly intellectual character of many of those so-called displays of intellect amongst animals. Take, for example, the statement that ants ascend the stalks of grasses, gnaw off the grains while others below detach the seed from the chaff and carry it home: then, after having allowed it to germinate and so convert the starch into sugar, *i.e.* after the grain is malted, gnaw off the radicle to prevent further germination. Now these statements heard of for the first time (but not witnessed) have excited grave suspicions, not to say great contempt, as indicating so much intelligence that an *à priori* disbelief immediately arises: but when the same phenomena are witnessed, as lately by Mr Moggridge at Mentone[1], we then begin to suspect that it is a constant occurrence amongst certain ants: and we shall soon regard it as part of their regular habits, and it will then lapse into instinct as much as bees collecting saccharine

[1] *Harvesting-Ants and Trap-door Spiders.* Reeve and Co.

fluid which they convert into honey and store for a winter's use. No one regards that as reasoning now; nor, if the storing of seeds were seen to be as common as honey, should we regard it of a more intelligent character than an acquired hereditary habit, *i.e.* Instinct. It would, however, look like an act of intelligence in the first ants that did it.

To return to Man, the general conclusion is that whichever way we regard him, whether with reference to the body, or his intellectual or moral[1] capacities, the gap between Man and Troglodytes is so great, that in the present stage of our knowledge no known forces can account for it; and the result we are inevitably driven to is, that Man is separated by so great a distance from the rest of the Primates that, as far as his body is concerned, he requires to be placed in a distinct 'family,' as to his intellect, certainly in a distinct 'order,' and morally in a distinct 'kingdom,' which may deservedly be called the 'Kingdom of Heaven.'

[1] The cause of Man being the only *moral* creature will be seen at pp. 150, 151.

CHAPTER X.

THE ARGUMENT FROM ANALOGY.

A LARGE class of evidential matter has not been touched upon, or scarcely so, in this Essay. It may be called *The indirect evidence of Analogy.*

When any principles are found to be of wider application than bearing only upon some particular phenomenon under discussion, and to embrace or apply to all other phenomena which may be akin or analogous to it, then the value of those principles, whatever they may be, will be proportionally strengthened.

Mr H. Spencer (*First Principles*, p. 307 *seqq.*) has in his general theory of Evolution furnished us with an admirable *résumé* of illustrations of the principles of Evolution as applied *e.g.* to painting and other arts, as well as to civilization and language; the latter being so well discussed by Sir C. Lyell in his *Antiquity of Man* it would be presumptuous in the writer to do more than call attention to those learned authors: and to express his conviction of the extreme value of their analogies in supporting Evolution when applied to the Organic World.

The particular features, however, which may be specially alluded to, are the principles, if we may so call them, of *Integration* or consolidation coupled with that of *differentiation* of organs and functions.

In the organic world we have seen how in the lowest group or that of simplest structure, those functions which are performed by different organs in the higher animals, are executed in common by the general structure. In other words, the more complex in structure is the organism the more differentiated are the organs by which the several functions are performed.

As an illustration of these principles taken from the Animal Kingdom, we may consider that while a Mammal has a heart, lungs and digestive organs, the sea anemone, which also has a flow of nutritive fluid, breathes and devours, yet effects these processes by one and the same simple organ or sac.

Now, these principles when applied to communities are well illustrated by 'Division of Labour,' which is an elaborate differentiation having its origin conjointly with integration of the social elements. In the savage or undifferentiated state of society each unit or individual meets his wants out of his own unassisted powers. On the other hand, in the highly civilized community, while we see the units aggregated together to make a complex whole, or an entire nationality, yet each unit has now its own proper duties to perform—the several duties of the units conjointly conspiring to maintain the general welfare of the whole.

Again, let us consider Music. In its simplest form there is only a monotonous tone—or a short cadence of a minimum of notes. But with these simple elements of short cadences combined, see what elaborate integrations, 'compositions' as they are rightly called, we have now obtained. But more than this—while music has differentiated, instruments have also; and though each instrument can be played alone, yet witness the supreme integration in an orchestral band: each instrument having its own part to play, yet contributing to the harmonious unity of the whole.

Amongst these analogies that which strikes one as having an importance far surpassing all the others put together in its ultimate consequence is the inestimable value which this principle offers to support Christianity.

The present writer has elsewhere[1] maintained the *naturalness* of Christianity to be one of its most important evidences, and on the present occasion he thinks it not out of place—but rather the reverse—to point out the application of this principle.

---

[1] Unpublished Sermons: as well as in a paper read before the Victoria Institute, on *Certain Analogies between the methods of Deity in Nature and Revelation.* (*Trans. of Victoria Ins.* Vol. IV.)

In St Paul's first epistle to the Corinthians, the great Apostle is reasoning upon the existence of a diversity of gifts, but that it was One and the same Spirit who divided severally as He willed.

He does so in consequence of a great amount of uncharitableness and bitterness which had arisen amongst the members of that Church, because certain of them had been endowed with gifts which others had not: and the great duty which St Paul tried to impress upon them, in the 13th chapter of his first epistle, is that of mutual love or Charity. For it was not a principle of God's Church, any more than in nature, to allow units to be all alike, and so independent and without cohesion: but that the different elements should be differently endowed, one element thereby supplementing by one kind of gift the wants of another, and *vice versâ;* and that all the diverse elements should harmoniously unite as one compact body: so that with 'differentiation' combined with 'integration' was the Church to be constructed. St Paul illustrates this principle by referring to the structure of the human body, which of course is a natural circumstance substantiating this great law.

Here, then, we recognize a great and fundamental principle, linking together in one scheme all creation both material and spiritual, and equally applicable to the individual units in the development of their structure and functions as to the collective units constituting a body material, a body politic, or a body ecclesiastic.

If ever the grand and final result of integration take place, it will be when all nationalities shall have learnt the great duty of life; that while existing as separate peoples, there must be international unity and harmony with mutual co-operation;—just as physical conditions will. ever give rise to differentiation in Beings, together with its concomitant phenomenon, integration;—this, however, will probably never be fulfilled before the end of all things shall have come, and terrestrial imperfect humanity ceased to exist.

We may add to these principles of differentiation and integration that of Retention of Type or Persistence of Character, as well as the laws of Embryology, as being equally applicable to all other things capable of growth and development as to organic beings.

As an illustration of the principle of Embryology alluded to, Mr Darwin has described to us its application to Man, so that we need not reproduce that instance here (*Descent of Man*, Vol. I. p. 14 seqq.). As another, however, may be given the following.

"In a very early stage of its development, the heart of the chick appears as an elongated sac or contractile tube, connected behind with veins, and in front with an artery. Thus in its general condition it resembles the simplest form of heart met with amongst the invertebrata. Soon the tube becomes bent upon itself and divided by two constrictions into three compartments, freely communicating. The one in which the veins terminate becomes auricle, the next ventricle, and the third an arterial bulb. Now, in its general condition, it resembles the heart of a fish.

Then, the single auricle, ventricle and arterial bulb each becomes divided into two, so as to form the left and right heart, as it occurs in birds and mammalia[1]."

The principle herein involved may be expressed by saying that—at any given stage in the development of the higher being, it has attained a grade of development equal to the more or most advanced condition of the lower: but, save in the equality of the grade, the characters are more or less distinct.

Thus, as in the structure of the higher animals and in Man the ankylosed condition of the bony parts is a subsequent one to an unankylosed state, so in fishes we see the arrested stage of the unankylosed condition: though as far as the existence of fishes is concerned, that stage or grade is the best suited to their mode of life and is *their* final stage of differentiation.

Now this great principle is also universal in its application; and applies to the development of mental phenomena just as much as to organic

---

[1] *Life and Death*, p. 29, by W. S. Savory, F.R.S.

structures. This has been already alluded to (p. 118), and therefore I shall say no more than merely recall the fact to the reader's mind, viz. that, if we compare the manners and habits of savages, they will be seen to be on the same mental horizon with children of civilized parents. Physical prowess with both is the great subject for adoration, as far as the male sex is concerned. Though the love of self-adornment is more differentiated among civilized beings and is, or ought to be, confined to the female sex; we find it undifferentiated in the savage, for he still decks himself out with beads, feathers and earrings[1].

Now let us apply this principle to Religion. As long as the Jew, who had been 'selected' to be the keeper of the elements of true religion subsequently to be developed into Christianity, was in a semi-barbaric state, all that could be taught him profitably was mere *humanity*, to

[1] It may perhaps be hinted that these same two features o external adornment and prowess find another illustration in our military, whose profession of arms necessarily cultivates the latter feature, while the former is at least associated with it on natural principles.

some slightly greater extent than was recognized by contemporary nations, amongst whom it was mainly ignored or only took the form of spasmodic instances of natural kindliness; a smouldering spark amongst the decaying remnants of natural morality, which was not to be quenched, but subsequently fanned into an ever-burning flame of Charity or universal love. Coupled with those lessons of humanity were numerous laws civil and moral to be kept, commandments, the chief spirit of which was "Thou shalt not—" with the promise of tangible rewards for obedience, such as an increase of corn, wine and oil; with temporal and even corporeal punishment for disobedience, the sword, fire and the plague.

Now what is the general bearing or interpretation of all this? That "the Law was a schoolmaster to bring [man] to Christ," that is, the Jew of the Old Testament represents Man in his *boyhood*, and the whole principle of discipline throughout is applicable to that stage of progress; but as soon as he becomes a man, and is "no longer under tutelage"..."he putteth away childish things." He is now, *i.e.* under Christ, "an heir, joint-heir with Him of the inheritance prepared for him of his Father."

An important remark must be added here; for this analogy of Evolution as applied to religion fails at one point.

In the development of art, science, or society, &c. we have a natural process working out the gradual integration with differentiation similar to division of labour; but with Christianity it is not entirely a natural process of development, but a *regeneration.*

The element of moral evil or sin has become a deterring agent in the process. And just as in the case of disease which must be cured in the body, Nature works at her reparative processes but cannot always effect a cure without external aid: so in the case of Religion, though the development to a large extent is natural, or at least in accordance with natural principles, yet humanity could not be restored, nor Christianity developed, by natural processes alone.

God interfered, therefore, like a physician, to

help Nature, to furnish an impulse to her powers of restoration, which had become enfeebled if not ruined! And, too, just as a cure cannot be effected all at once but is usually a slow process, so in the regeneration of mankind there have been many divine interferences—in 'selecting' one man, Abraham; then in the protection, guidance and training of the 'peculiar people,' Israel; while the greatest and most important interference was when Christ the Son of God, the Physician of the soul, came and promulgated the fundamental conditions of universal love to all mankind, as all men are brothers by nature: while at the same time the basis of self-regeneration must be laid in an enthusiasm roused by a love to Him.

In other words, by no natural process of Evolution could Christianity be differentiated out of Heathen philosophy or out of the highest phase of Judaism—Pharisaism—at the period of Christ's coming: and no other phase of religious or moral thought existed at the time[1].

[1] I would refer the reader to a most valuable work on this subject, *The Jesus of the Evangelists*, by the Rev. C. A. Row.

But, here, once more mark an analogy to Nature even in this divine interference of Christ.

Just as man agrees with animals in all fundamental points of structure, even to the possession of rudimentary organs, as well as mental qualities, though moral and spiritual features prove him to be distinct, yet in the aggregate his powers are far in advance of theirs; in fact, in every way, as we have seen, is the gap so enormous, that no principles that we are acquainted with will justify us in saying that he could be evolved from the *Simiadæ* without some divine impulse being given to him.

So when we compare the old ethnic ideas of what was noble, with the teaching of Christ; who exalts the moral and humbler virtues over the purely physical and heroic ones; and who has thrown a sacred halo over the most degraded and abandoned of mankind, whom the philosopher admitted he was powerless to regenerate by any system of morality or philosophy he could propound, and whom the proud Pharisee could only pronounce as 'cursed'—when we think of all this, what Christianity *is*, and what it has done, the analogy is forced upon us that the gap between it and ancient philosophy is so enormous, as to preclude the idea of its ever having been developed out of any phase of moral thought which existed at the time, when we know from testamentary evidence that Christ came and propounded His doctrines of universal love and everlasting life through Himself.

Another analogy is yet worthy of note.

No one can fail to be struck with the remarkable parallel between the history of animal and vegetal life and that of nations. If we compare those of Eastern countries or semi-barbaric ones with those of Greece, Rome, or England, we see the same lengthened continuity of an uneventful stream for ages, whereas in a highly and rapidly developing nation—*i.e.* that undergoing differentiation—thousands of events, political, social ecclesiastical, or what not, mark the line of progress in as many years as there were of uneventful centuries in a state of low civilization[1]. Every feature remains long undifferentiated or

[1] The above idea is borrowed from Prof. Haughton's *Manual of Geology*.

stereotyped; art becomes stagnant and runs along permanent grooves without any, or with but slight, modifications. Beliefs and superstitions become fossilized, so that the feature of Retention of Type (and the lower the grade the more persistent is it) holds good for all other classes of phenomena as well as for the Evolution of life.

What may be considered still more striking is that similar states of barbarism have analogous if not absolutely identical characteristics. Nothing could be more illustrative of this than the universally similar types of flint and other weapons of savages; thus the chipped flints of Europe are paralleled by the chipped obsidian of Mexico. The smooth and polished celts of the old Continent find their counterparts in the polished jade of the New Zealander of to-day, illustrating well the relative persistency of whatever is of a low grade[1].

[1] A few illustrations of a retention of type are the following, *viz.* the uniformity of the plays of Terence compared with the great variety of the modern drama: the characters of existing Chinese and Japanese art: the *inlaid pattern* of Indian ivory ornaments and boxes being always the same: lastly, the clown of a pantomime, the letters of rustics and the present type of four-wheeled cabs.

One more feature will be noticed and then this chapter on analogies must close.

It may be quite true, as Mr H. Spencer says, that there is a passage in every 'Life' "from the imperceptible into the perceptible and again from the perceptible into the imperceptible;" but he does not bring out what appears to be another important principle, *viz.*, that during the passage into the imperceptible, "A remnant shall be left which shall take root downwards and bear fruit upwards" (Is. xxxvii. 31), not necessarily to resume the diagnostic features of the Being when perceptible, but to be differentiated or at least so modified afresh as to appear as a new specific entity. Though the principle is quoted from the history of the Israelites, *yet it seems equally true of all beings*[1].

---

[1] Since writing the above, the following observations by Mr G. Bentham have appeared in the *Journal of the Linn. Soc.*, Vol. XIII. p. 481, which bear a remarkable corroboration of this idea. Speaking of the dispersion of races of the *Compositæ*, and of the origin, the rise to full vigour and final decay of any group, he says: "Old, decaying and apparently expiring races may in some of their branches...start into new life." "Young progressive races...may be rising before our eyes from some branch of an old race which has passed its prime." Again, with regard to certain

Of organic beings, as we have seen, each type presumably commenced with an insignificant origin, rose to perceptibility and characteristic significance, then disappeared, leaving a remnant which pointed to its former existence, while new types were differentiated out of those remaining.

Thus has it been with Christianity; the germ is seen in Abraham, the family in Jacob, the tribes with the Judges, the nation with Solomon. The first step towards dissolution appears in the contest between his son Rehoboam and Jeroboam. The ultimate collapse is witnessed in the dispersion; but a remnant did return, struck root, and out of the seed of Jesse rose the Righteous Branch which flourished, and bears fruit amongst us now.

Just so, too, do other nations, political communities, sects and other bodies, rise, flourish and decay, but then the remnants give rise to new combinations, and reappear in fresh forms ever gathering strength to become new specific or

"races still vigorous but breaking up into subordinate races, many...give indications of future diminution and extinction, but some of which as yet *of a very low grade* exhibit a great susceptibility of extension and progress."

generic types of being, till at last they too have run their course and disappear, and so the cycle is completed: and thus will it be with all things until the end of time.

Thus, then, do we see how all those groups of phenomena—which, when brought to bear upon the history of animal and vegetal life, *force* upon us the acceptation of Evolution, as best accounting for their origin, rise and disappearance—apply to every class of terrestrial 'being,' organic, mental or spiritual, which is characterized by possessing powers of growth, development, or differentiation.

The immense value of this discovery lies in the disclosure of grand uniform principles, as governing the development and differentiation of the Universe, thereby testifying to the unity of design throughout and the One Final Cause of all.

A fact of great significance with regard to the harmony between Christianity and natural law is that Christ so frequently illustrated the development of His kingdom by similes taken from natural phenomena. His parables are often founded upon nature; thus the law of slow development of religion, whether in the individual or the nation, is likened to 'the mustard' or 'good seed;' while the 'struggle for existence' of the Word in the heart, against bigotry, ignorance, or the superstitions of a dominant power, is illustrated by the seed growing by the way side or amongst thorns.

The reader may here be reminded that the effects of Christ's 'fan,' at the commencement of His ministry, were as true an instance of Natural Selection as any in Nature; for it was only those hearts which would receive His teaching, that were thus self-selected as His disciples, the 'fittest survived,' the proud and self-conceited falling away into prejudice and unbelief.

It must be noticed too, how in Religion the same principles hold good for the individual as for the nation. Just as the history of a man's life is an epitome of that of a community, or as a human embryo epitomizes the rest of the mammalia; so does the growth in grace of a man symbolize the spiritual advancement of a nation.

CHAPTER XI.

MORAL EVIL OR SELFISHNESS.

IN the last chapter I reminded the reader, that Christianity is not so much a development as a regeneration of mankind; and it will not be out of place to give some account of the cause of the moral degradation of man, which has required such a regeneration.

Much has been written concerning moral evil, 'the Fall' and Sin: and though it may seem somewhat superfluous to enter upon a disquisition of these subjects here, yet as they have their collateral bearings upon Development, especially in its theological aspects, or Christianity, it is desirable to make some few remarks upon them.

I would state therefore that my belief is, that the origin of Moral Evil was *the conscious abuse of means instead of using them solely for the ends for which they were designed*[1].

In order to see the full force of this assertion, we must compare the condition of animals in their *feral* state with that of Man.

Now, I think it is a fair statement, to say that such animals as a rule do no more than follow their natural instincts.

In obeying those laws of self-preservation and propagation, which have been impressed upon them, it is extremely probable that wild animals eat and drink, not for the sake of eating and drinking, but to maintain bodily life only. The laws of propagation are periodically in force, and are obeyed at those times; but union is

---

[1] I have just met with the following remarks in Dr Goulburn's *Thoughts on Personal Religion*, p. 300, and gladly subjoin them in a note, for they seem to correspond exactly with this view: "So apt are we (at least in spiritual subjects) to confound means with ends and to erect the means into an end—an intellectual perplexity, indeed, but one which sometimes appears to me to bear a trace of the Fall, and to be due ultimately to the corruption of our nature."

probably not resorted to, for mere union's sake. In fact animals, as far as we can judge, shew no signs whatever of distinguishing the object from the means; and they are consequently not conscious of the possibility of abusing the means instead of using them.

Now with man there is an element of consciousness which, even if it could be detected in animals, certainly so far transcends the extent to which it may be developed in them, as to bring about vastly different results. I mean that he alone can *see* that eating is pleasant, and so often eats for the sake of eating: and similarly for other pleasures. This appears to me to be the right interpretation of such an expression in Genesis as the "knowledge of good and evil," that is to say a consciousness of the right use of natural law and the 'evil' which lies in the abuse of such good gifts as the means by which the natural law is executed. That moral evil or sin takes its rise in this abuse of means, or lust, which is the same thing, is supported by various expressions in the New

Testament: thus St James says, "Every man is tempted, when he is drawn away of his own lust and enticed."

If now I am right in assuming the first members of the human race to have been under low moral and intellectual conditions, the first element of 'sin' would be this kind of departure from rectitude. For while, as we now know, consciousness of probity, when developed into the form of that of the highest civilized Christian, places him at an immeasurable distance above animals, barbarians and savages; yet that very consciousness, where the intellectual and moral senses are weak, renders man an easy prey to those enticements which lead him to abuse Nature's gifts instead of using them. Lust is the first element of trial to which a man can be subjected.

As human beings multiplied and formed social masses, and as civilization advanced, evil would become 'differentiated;' it would take, and has taken, diverse forms. Thus when coinage was introduced and money was looked upon as the representative of goods, and as a means for an end, then the love of money *per se*, as exhibited for example in the miser, represented an abuse of the means to the sacrifice of the end. But not only is it the miser but the selfish squanderer who sins. The use of money is purely relative, and after a just proportion is spent upon oneself, the surplus is not to be squandered but bestowed where it is most needed.

"Let him that stole steal no more, but rather let him labour with his hands, *that he may have to give* to him that needeth."

So too Christ says, "Make to yourselves friends of the Mammon of unrighteousness, that when ye fail they may receive you into everlasting habitations."

So that whether a man hoards or whether he squanders, the sin is the same, for it involves a gross abuse of wealth, which is only a means to an end and not the 'end' itself.

Another instance is lying: this is the abuse of speech for selfish and wrong ends, and is a secondary result from 'knowing good and evil.' See James i. 26 and iii.

Now it will be observed how all classes of sin centre in selfishness. With regard to sins of the flesh this is obviously so: with lying it is an attempt either to obtain the selfish end of securing oneself from injury, or to gain some supposed wish or gratification, as in exaggeration, romancing or scandal. With stealing, it is the endeavour to obtain for oneself the belongings of another. With passion, malice, and murder, it is the gratification of selfish feelings of revenge.

Now in these, and the reader may perhaps suggest others to himself, a process of development or differentiation of evil or sin may be observed to have taken place.

Thus in the case of anger.

There is a righteous indignation or anger, as it is called in Scripture, when it is said that "Jesus looked round about him in anger," or again, "He was angered at the hardness of their hearts." In this there is no sin; it is a natural instinct like conscience; for indignation detects or upbraids the sin of others, while the conscience upbraids sin in ourselves.

But when this kind of anger is cherished for anger's sake; when we have "let the sun go down upon our wrath," it assumes the forms of rage, hatred, malice and revenge. These are its differentiations, and are relatively worse and worse, till the lust or selfishness of our rage culminates in the crime of murder.

Or again stealing, and we may add begging. Disregarding very exceptional cases of starvation when natural cravings overmaster the moral sense, stealing and begging may be regarded as forms of selfishness, at least when resulting from idleness.

Idleness is characteristic of young days. When the infant, having had everything done for it, becomes a boy, the being put to a task to be undertaken on his own account appears irksome to him, and idleness is a not unnatural feeling; then, if this be not thrown off, but remain as part of the constitution of the man, the vicious results are but too obvious. It is an arrested stage of man's development. Idleness in a man is equivalent to his remaining like a boy in habits. Whichever way, then, we look at sin, it lapses into selfishness. "The lust of the flesh, the lust of the eye, and the pride of life" are but various forms of self-gratification.

Hence the word *selfishness* is synonymous with *sin*; and it becomes us to consider how this selfishness has arisen.

When we examine the habits of animals, self-interest is undoubtedly a ruling power.

By self-interest in animals I do not mean the same thing as the *conscious* self-interest of a man; but that animals solely follow the guiding of the laws which induce them to acquire food for themselves and to preserve themselves from injury. That is to say, all their actions are instinctively directed to their own good. This self-interest of animals is purely instinctive, and involves no *consciousness* whatever. This same remark applies to the apparent selfishness and unselfishness of animals.

Indeed the word selfishness is purely relative, and can only be applied where the means have been consciously abused.

Acts apparently unselfish, not to say of a charitable appearance, have been witnessed in animals; as when a blind member of a community has been fed and cared for by its kin; or again when an animal has risked its life for another, or for its keeper. Still, in these and all such like cases, there could be no consciousness of the self-sacrifice, certainly not of any duty to a Higher Power.

Here then we may note the grand distinction between man and animals lies in this *consciousness* of the nature of such acts as these and the sense of *duty* involved in them. With animals, acts of unselfishness are without consciousness of duty. Hence we see why man alone can be *moral*. With man unselfishness has to become a leading principle of action or duty.

Self-interest with man beyond a certain radius round necessity becomes selfishness, and is culpable according to its degree, as means will then become abused.

The great law of unselfishness or self-sacrifice is the basis of natural religion of man as it is of Christianity, but could only have been implanted in his nature by God.

To be unselfish was, I have no doubt, the original and normal condition of man, when he first became a "*living* soul;" but the abuse of means has brought about the selfish state; and as generation after generation has passed and gone, selfish habits have been copied and inherited until selfishness replaced liberality, exclusiveness overthrew generosity, and hostility was one and the same with alienation. Sin became the recognised and supposed right condition and so was unknown as sin; nay more, it became part of religion itself; "the prophets prophesied falsely, and the people loved to have it so." In fact "conscience had taken the wrong side."

We may observe, too, how selfishness is a disintegrating process, and is the very reverse of Evolution in its effects.

The Evolution of societies, as we have seen, is the gradual differentiation of their elements, which, however, remain linked together by mutual co-operation, as for example in the principle of the division of labour. Amongst savages on the contrary there is no cohesion. Each has to supply all his own wants, and is at enmity with all around.

So too the selfishness and exclusiveness of clans, together with their hereditary hatred towards each other, are the means of destroying the power of Evolution and disintegrating its effects. On the other hand the slow fusion of clan with clan in the bonds of friendship, the slower binding of nations by international unions, are results of the Evolution of society, or of the gradual leavening process of Christianity, for in this they are one and the same.

Another aspect of selfishness:—

Every man stands in the midst of his fellowmen, and his mission consists in doing as much good to them as possible within his little sphere of action. He himself is a *means* to an *end*, "in honour preferring another" rather than himself. If he neglects that duty, wraps himself up in a cloak of selfishness, and shuts up his bowels of compassion, he is abusing that 'means' and sacrificing the chief object of his existence for his own gratification.

The social nexus of Christianity and of Evolution is one and the same, namely unselfishness or self-sacrifice.

To repeat in summary the general effects of selfishness is to say that its tendency is to make men isolated or disintegrated beings, and is therefore the very reverse of the laws of Evolution, which are well expressed by Mr Spencer in the words "differentiation with integration," whereas disintegration crumbles all to pieces, and makes men isolated and *repulsive* units.

It is easy to see, then, that just so far as selfishness holds ground in individuals or in a party, clique, religious sect or what not, to that same degree is there disintegration in the general body to which they belong; and that the philosophy of Christianity and the philosophy of nature, *i.e.* Evolution, encourages and develops, if it cannot at once enforce, the exactly opposite conditions.

That the stability of the original naturalness—the golden age of the ancients—has been overthrown, and that the moral equilibrium of man has received a shock of continually increasing impetus through the ages of the world, the pages of universal history too plainly disclose.

Nothing more clearly proves this than the condition of the world from its earlier times down to the present, and especially when under the influence of Christianity. War was the normal or usual condition; "not," as the author of *Ecce Homo* says, "because there had been any quarrel, but because there had been no treaty." The very word *hostis* was equivalent to 'foreigner' and 'enemy' alike. But the condition which appeals to every civilized man's heart as most conducive to happiness and comfort is universal peace—not universal war. Every day, as civilization advances under the laws of Evolution, war becomes more disastrous to the community, which in proportion as it realizes the temporal advantages of peace, so does it dread the necessity of war: and what therefore one can easily see to be the *summum bonum* of national prosperity, namely, international unity, is the grand aim of civilization, Christianity, Evolution, and humanity.

This is the grandest aim of the Gospel, and was uttered in the words of the prophet Isaiah when he proclaimed the time that men should "beat their swords into ploughshares and their spears into pruning-hooks:" and that "nation shall not lift up sword against nation, neither shall they learn war any more." Is. ii. 4.

Christianity, then, is the religious form of Evolution; though in consequence of the accumulated degradation of man, the impetus of evil had acquired too great power to be reversed by any natural forces in existence. Badness was thought to be goodness. The natural instincts of purity, of kindliness, of mercy, charity, self-sacrifice and others of which the Gospel of Christ speaks, had become long since crushed down, ignored and forgotten, so that no purely human effort could drag forth and elevate them so as to become ruling powers over the dominant forces.

But, however, having been at last by superhuman, nay, Divine means, brought to light, and the degrading effects of sin, now recognised as *sin*, put forth, the laws of Christian Evolution have once more gained a footing. The seed of the Tree of Life has been sown afresh, its fruit is rapidly spreading, and let us hope that, like a vigorous alien amongst plants, it will rapidly drive out the aboriginal weeds of idolatrous and debasing superstitions.

CHAPTER XII.

ON THE WISDOM AND BENEFICENCE OF THE CREATOR.

HAVING seen that *man* has appeared as the last of the Creator's works; and that, although linked with other vertebrates, yet can he boast of far higher powers—intellectual and moral—than any other creatures; it is time to consider the terrestrial history of life with reference to the Creator Himself, and endeavour as far as possible to discover the wisdom and beneficence of God as displayed in the Evolution of Living Things.

In endeavouring to trace evidences of the wisdom and beneficence of God in the Evolution of Living Things, certain difficulties will soon become apparent: and it is necessary at the outset to understand clearly what we signify by these attributes of the Deity.

We cannot judge of actual wisdom beyond our own experience and consciousness, and when we apply the word to man, we signify "a right use of knowledge;" when, however, we try to estimate the wisdom of God, we fail from our feeble powers to grasp so transcendent an attribute as it becomes when vested in Deity.

It is easy enough to detect thousands of what we may designate instances of wisdom in the details of creation, as, for example, in supplying every kind of creature with a proper means of obtaining its food and enjoying its existence. Wisdom is obvious in furnishing every animal with organs suited to its sphere of existence and so forth: or, again, in the phenomena of anatomy, the combinations of strength and lightness, rapidity of muscular action with no loss of force, or what is called 'the principle of least action[1].'

[1] See Lectures on this subject by Rev. Prof. Haughton, F.R.S., delivered at the Royal Institution.

Now it must be observed here, that works on Natural Theology do not appear to go beyond this; and that the writers, such as Paley, aim no higher than to shew that such wonderfully constructed organs as exist in the human body must have had an intelligent Designer—arguing of course from man's powers of designing and constructing.

Or if one step further is attempted, *viz.* to shew forth the Divine wisdom and goodness, some writers, as of the *Bridgewater Treatises*, do not, as a rule, seem to see that all instances of goodness and wisdom are *relative* only; and therefore not unfrequently they are guilty of penning the most flagrant instances of false or questionable reasoning[1]. Merely, therefore, to bring forward instances in any degree of abundance (as could easily be done) of wisdom and beneficence, *which might at least be considered such by one being in respect of itself only,*

---

[1] See the writer's paper on *Natural Theology, considered with reference to Modern Philosophy.—Trans. of the Victoria Institute,* Vol. VII.

is a waste of time. Such may be considered an axiom of the methods of Creation.

But when those same instances of goodness and beneficence are regarded, from the point of sight of another being, the latter may not consider them beneficial at all. Thus, an Ichneumon fly, had it reasoning powers, might easily conclude that caterpillars were beneficently designed for its use, as being the place in which it should lay its eggs. On the other hand, the caterpillar would have a very different view of the beneficence of the Being who made both itself and the Ichneumon.

I repeat, therefore, to bring forward instances of advantages or even blessings to one being, even if we select Man, as proofs of the wisdom and beneficence of God, is, and must be, a partial and one-sided view of Creation: and this, if I understand them aright, is an unmistakeably weak point in the arguments of the writers of the *Bridgewater Treatises*, and of Natural Theologians in general.

The present writer, therefore, will bring forward but few of such instances, or perhaps none, as proofs of wisdom and beneficence; for, as has been already stated, they are self-evident when only a personal or selfish view is assumed.

On the other hand, it will be the object of this Essay to discover God's wisdom and beneficence, as far as Man's limited powers *can* trace it, in Creation as a whole: or rather in those great principles which govern the evolution and development of all Beings in the world.

The reader must, therefore, bear in mind that any observations which seem to shew direct instances of wisdom and beneficence towards Creatures, must not be isolated from the chain or circle of physical phenomena as well as from other beings with which they are intimately connected. Such observations will ever be of a limited, partial and inaccurate character. And we must remember that besides this system of dependent causes and effects, there still remains the far deeper question, Why have animals ever existed at all?

We may add to this such a question as the

following, Why is it necessary, if we admit some, that others should prey upon them, and so cause a certain amount of pain, rather than that the power of increase in those preyed upon be limited? Whichever way we turn to find indications of wisdom, we find that they are only contingent upon other circumstances, which raise a fresh question as to the wisdom of *their* existence.

Hence, what we may deem an instance of wisdom in one case is so relatively to some other conditions which exist, and these must be accounted for and their wisdom proved.

What, then, is the final result we come to? Simply this. The *Wisdom* of God as displayed in the works of Creation is synonymous with the *Will* of God.

Now this is true in the broadest sense, as far as the creation of everything both in time and space is concerned, including Man himself. Or, if we first say, that it was the will of God to create a moral being, Man; then, we can somewhat estimate the wisdom of the existing relationships among the different orders of beings in the world so far as they bear upon Man's existence and hopes.

In endeavouring to trace the wisdom of God we *must* view it in this latter sense only, that is to say—by making it contingent upon some one instance of the exercise of His Omnipotent Will.

Man's limited insight into the methods and purposes of the Deity compel him to come to this conclusion. It may be very like cutting the Gordian knot, but when once we get beyond positive evidence and try to investigate causes and motives we are attempting to escape beyond the confines of the human intellect. Not that man may not try to penetrate some way into the labyrinth of truth and track out the clue to the discovery of her Final Cause; but he will probably never succeed in estimating aright in every instance the secret purposes of God until the time shall come when "we shall know as we are known[1]."

---

[1] Several instances of such false estimate of purposes could be cited. (See pp. 12 and 13, in the paper on *Natural*

Recognising, then, the fact that the wisdom of God in nature is synonymous with the will of God, we will isolate one act of creation alone as representing God's will, *viz.* that of the Creation of Man.

The purpose of this Essay will henceforth be to shew the wisdom and beneficence of God as exhibited in Evolution and its consequences, especially in its bearings upon that assumption.

If now we proceed to search for proofs of the Beneficence of the Creator as distinct from His Wisdom, we shall find ourselves apparently involved in even greater difficulties than in tracing evidences of Divine wisdom.

We can only reason from man's estimate of beneficence;—which, apart from revelation, takes no higher form than *natural kindliness;*— and judging by that, the evidences of it from nature may be fairly questioned: while apparent proof of the very opposite can quite as, if not more, readily be produced.

Theology considered with reference to Modern Philosophy, by the writer, *Trans. of the Victoria Institute*, Vol. VII.)

Indeed, it will be remembered that Lucretius of old declared that the world could not have been created by the gods or for their pleasure, in consequence of the great amount of physical evil present in it. We may notice also that the existence of physical evils, which ranges over every class of natural phenomena that concerns living beings, has ever been the perplexing feature in nature, not merely to the Democrital but all other philosophies of every age; and still is the basis of one of the arguments of the modern Atheist[1].

If, now, we are required to shew evidence of the beneficence of God as evinced by the Evolution of Living Things, it is very difficult if not absolutely impossible to find a satisfactory answer from the works of Creation alone.

---

[1] The present writer was requested (*March* 12, 1872) to deliver a lecture on 'The Argument of Design,' in Mr. Bradlaugh's Hall of Science; and in the discussion which ensued with professing Atheists (Secularists) he was most forcibly struck with the intense feeling they exhibited with reference to physical evils; realizing to his mind, more fully than anything else could have done, the sad hopelessness of a man who looks for no future beyond the grave.

If, for example, we try to estimate the amount of comfort against discomfort, pleasure against pain, happiness against unhappiness, as shewn by the animal creation; even if it be granted that there is a preponderance of the former, yet we cannot deny that the quantity of the latter nullifies all idea of absolute beneficence: presuming it to have been all the work of the same Almighty Creator.

We may reasonably ask, If God could *will*, and had infinite power to *execute*, why does He permit the existence of physical evil at all? Why not make a world exhibiting absolute beneficence in every, the minutest detail, both for animals and man?

Now, no study of external nature alone has ever offered a satisfactory answer to such questions.

A few reflections, such as the above, soon convince us that the words goodness and beneficence can only be considered from a relative point of view. They are doubtless absolute qualities or attributes of God's own Holy Essence; but in their application to Living Things a careful scrutiny of the conditions of existence of all kinds of life cannot possibly bring out more than a relative state of happiness or comfort.

Let us consider this more in detail: and the first question which naturally arises is, Why have animals existed at all? or, What is the object or design of life? Limiting the enquiry to the animal creation, exclusive of Man, and endeavouring to estimate what advantages any animal may appear to gain by living at all, a palpable answer in favor of some decided advantage is *not* forthcoming. Seeking for food, eating until satisfied, sleeping, with some sense of enjoyment when free from terror of an enemy or especially at the period of pairing, when pleasure of a social kind is tolerably evident, make up the average existence of an animal. Its prey is not so easily obtained if it be carnivorous, but that it must be wary and watchful; nor is its own chance of escape from an enemy so perfect but that it may sooner or later fall a victim. It is very difficult to say positively that its total amount of enjoyment more than compensates for its total amount of discomfort, fear, or physical pain. And, if it be the work of a good Creator, our ideas of absolute beneficence at once lead us to ask, Why is not its existence one of perpetual and absolute enjoyment without the possibility of pain, whether of body or mind[1]?

Moreover, animals appear to be utterly unconscious of their existence—we cannot conceive any animal reflecting as to why it exists. It simply obeys impulses from within and without: it acts unconsciously by instinct: it may boast of a certain amount of reason, but its actions are nearly always regulated by law. Each kind of animal has its own peculiar prey. The same kinds of the predaceous seek for the same kinds of food and always in the same way. Each kind frequents the same sort of habitat; thus, insects lay their eggs instinctively on such kinds of plants best suited to the caterpillars which shall issue from them. So too as far as fun and

[1] The idea of suffering amongst wild animals having been due to 'the fall of man' must now be entirely abandoned, as without a shadow of foundation either from Scripture or science.

amusement are appreciated, they always play in the same way, each animal after its kind[1]. So that in point of fact all the specialities of the existence of animals are regulated by laws.

It will not be devoid of interest to repeat here the differences between an animal which obeys laws without reflecting upon them as such, and Man who *can* do so and uses his free will to obey them or not, in order to introduce an important principle of Nature.

An animal in a state of nature eats and drinks when it is hungry and thirsty, and does so until it is satisfied: but does not eat merely because eating is pleasurable. It has no power to reflect or to be conscious of the fact that eating is pleasant and a "thing to be desired," so never eats for eating's sake; whereas man, who knows "good and evil," that is, can recognize the purpose of these natural laws and that they are *good*, has distinctly abused them, and

---

[1] Ex. During at least the author's life, there has not been the slightest change in the habits of flies which dart backwards and forwards in a zigzag manner under some pendent object in a room, pirouetting whenever they meet!

so committed *evil*, simply because he can reflect upon the pleasure as a thing apart from the object of the law: he has indulged that pleasure when the law had no need to be executed; has exalted the 'means' at the expense of the 'end'; has (and herein is involved the principle) not only so vitiated his own nature by breaking down the efficiency of natural law, but has imparted the irregularity to his descendants in accordance with the well-known physiological law of hereditary acquirements, and which may be expressed by the words "the sins of the fathers shall be visited upon the children."

Again, the sexual appetites are in animals periodically regulated by law. Animals obey that law at stated periods. They do not break through it from any conscious idea of pleasure *per se;* and so never indulge the appetite with any such motive apart from obeying the natural law[1].

[1] This is certainly not without exceptions; yet, on the one hand, the general obedience to the natural law is well-nigh universal, and on the other, animals have no consciousness or idea of law at all.

Man has done so, and with a like result.

Such I believe to be the Origin of Moral Evil.

Lastly, we have no reason for believing that animals possess the merest shadow of a suspicion of the possibility of a future existence, of the being of a God, or of duty to any Superior Being at all.

Now, knowing as we do what pleasure is, and what pain is, we can conceive of a creation of animals capable of enjoying a far greater degree of pleasure than any which exists on this earth; and if we assume an Omnipotent and Beneficent Creator to have by His own will called the animals into existence which inhabit this world, we cannot but think that their short sphere of life might have been rendered absolutely happy, or at least far happier than we have any reason for supposing them to be.

Our actual ideas, therefore, of the beneficence of the Deity, when acquired solely from the observations of the habits and general state of existence of animals, become apparently much restricted.

To return to the questions, Why do animals exist at all, and what is the object and design of life? We have seen that a scrutiny of the lives of animals supplies us with no satisfactory answer at all. For—what has been repeatedly urged by Atheists and Sceptics—as much might be said against, as for, beneficence in the works of Creation.

But, as soon as we take Man into consideration, the light begins to dawn, and we can at once trace some clue to the interpretation of the question of their existence. But even now we must first find an answer to the higher question, Why am I here? for it is in the reply to this that the former question finds its response.

There is but one reply, which is, that God so willed it. This, as before, cuts the Gordian knot, but we cannot find any other satisfactory answer from nature[1].

---

[1] I repeat, Man cannot have been a necessary result of Evolution. The gap between him and all animals is far too great, not-

Moreover, notwithstanding their advantages, when we compare the state of existence of human beings with that of animals, it might be reasonably asserted that the latter—under such limitations as they are—seem often far better off than a vast number of our race; for it is the few, not the many of us, that succeed in having an abundance of this world's comforts. Hence, presuming no future or better prospects are to be expected after death, constant personal happiness must be looked upon as the ideal acme of conceivable bliss.

To interpret, then, the object of man's existence, we *must* turn to Revelation; and we are at once completely satisfied; for no answer equals this:—it was the will of God that there should be a being who could be moral, and that he should pass through a period of probation, before he

withstanding his bodily and intellectual affinities, to be accounted for by any laws of development that we know of; nor can we conceive of any force capable of being differentiated into the *Will*, a power which may be in direct opposition to forces of nature. We are then justified in concluding some special circumstances to have obtained in the first production of a human being. [See *Genesis and Geology, a Sermon*, by the writer, p. 20. (*Hardewick*.)]

be fitted to enjoy that state to which his spiritual part is naturally best adapted.

But this explanation, this key to unlock the mysteries of all life, cannot be satisfactorily nor easily found by the study of nature alone.

Human consciousness may have been intuitively persuaded of the truth of a 'Redeemer who would stand at the latter day upon the earth,' and who would set all things right; but it is the rare instance of a soul self-conscious of its own *natural integrity* appealing to the Author of its existence, trusting confidently in His wisdom and justice, and not doubting His beneficence even when writhing under extreme physical torture.

We must bear in mind, too, that, as far as all animals are concerned, we cannot say that physical evils are necessarily, if at all, of a disciplinary character; that, as a parent chastises his child with feelings of love infused into his sense of just retribution, so God does the same in nature. This may be the interpretation which natural religion would give us with reference to Man; but the mere study of animal life irrespective of him—for physical evils affect both alike—does *not* furnish us with any such reply; simply because we have no grounds for believing any animal to be *moral*, or to have a sense of *duty*.

But with Man it is very different. Physical evils, which cover the range of all classes of circumstances, are of the nature of discipline. Had man never fallen, and so vitiated his right view of things, he would no more have regarded them as 'evils' than do the beasts of the fields. But it is just because he is *now* conscious of evil as well as good, that they have become disciplinary; and in proportion to the degradation of his mind so will he be inclined to abuse his Maker for them, if not to "curse God and die." On the other hand, the humbler and purer a man's heart is, the clearer will he "see God"—in nature—and the less will he feel external evils to be really 'evils,' or that they partake of the character of discipline at all[1].

[1] I would most urgently desire the reader to bear in mind

But now let us return, and see how the question, Why do animals exist? finds its reply in that to the question, Why do I exist?

As has been already said, the vast majority of men in all ages could find no answer that was at all satisfactory; for they either looked for no future existence, or if they did, to one far different from that for which a modern Christian hopes.

We must clearly understand that *The future existence of man can be the only interpretation of his existence here.* The very ability he possesses to criticize his condition on earth and to imagine a far happier state, not to add an internal and conscious longing for another life,

---

the sharp line of demarcation between 'physical' and 'moral' evil with *its* results. I believe that much confusion has arisen from the close analogy between these two things, the one ranging over the physical world, the other over the moral; hence they have been supposed to be due to some common source. This, I believe to be a thoroughly erroneous idea. Physical 'evils' represent the principle of *Inideality*, as I would propose to call it, *i.e.* the fact that 'Relative Perfection' only exists, is part of the constitution of the world, and arises out of Laws of Evolution. Moral evil is due to the fact of man's *abusing* natural law, as stated above, and not *using* it only in obedience to natural impulse; this is what is called 'the fall' in Scriptural language.

is a witness of itself that he is at least by nature qualified to enjoy a better life, and has a right to consider that God would not have given him those yearnings if He did not intend to gratify them. On the contrary, if he do not believe in a God and a future state, the Atheist is bound to shew whence those feelings arise, out of what they have been developed, and what they imply.

They are qualities which alone separate Man essentially from all other creatures. Nevertheless the characters which link man to animals, viz. his bodily structure and his minor intellectual faculties, throw a great light upon the final cause of the existence of the animate (and through them of the vegetal) creation.

It is when we reflect how the organization of man is fundamentally the same in all essential particulars as that of every vertebrate, and how his appearance upon the stage of life did not take place until the world had in its previous acts displayed a grand succession of vertebrate life:—groups coming on, crossing the stage of existence, playing their part and then disappearing, ever and anon giving rise to new beings:—it is only when we reflect upon such a panoramic sight as this, and gaze down the interminable vista of the past, and find that man, the last member of the same troupe, came forward only at the very end of this long line of ancestry, that we learn how he has closed the series and revealed its final cause.

In regarding Man thus, as the last stroke in the great Creator's design of terrestrial life, it may be noticed again that, as far as his bodily structure at least is concerned, he certainly cannot be separated from the Primates; for it is in them that we see the last links of the chain of vertebrate life. Physiologists have also clearly proved that the same laws of development govern the human embryo as those of other animals; that is, it passes through representative stages of other vertebrates in an ascending scale. Moreover, man is not without rudimentary organs like other animals. Hence it will be seen that the *facts* upon which the doctrine of Evolution is based, when suggested to account for animal forms, apply necessarily to man's body, though exceptional features imply interference.

If Evolution be true for the former, it would seem that it must apply to man's body at least. Thus far, then, man cannot be severed from animals. So that we might almost say, in order to produce man it was necessary, to be consistent with the plan adopted by God, to evolve successively that long line of vertebrates from the Silurian Epoch until the present day[1].

Now, then, we may see how the popular expression, 'All things are created for man,' has a new meaning thrown into it. It is absolutely untrue and positively presumptuous in its current aspect. Indeed it will not be amiss to consider it, and point out the untenableness of that popular notion.

It is quite true that man cannot do without a certain or even great number of things—animals and plants for his food and clothing,

---

[1] See the writer's strictures on Mr. Lewes' expressions *Nature's bungling, Feeling her way*, &c. &c. with reference to embryology. *Natural Theology considered with reference to Modern Philosophy. Trans. of Victoria Institute*, Vol. VII. p. 16.

minerals for building and the uses in the arts, and so forth; but these clearly constitute but a small part of nature. Many things are at present useless to him, and many other things which he may desire, and even know where they are to be found, are inaccessible to him.

It is a poor remark to say, that everything is of some good to man if he did but know what the use was. This involves a contradiction. On the one hand, there is an instinctive feeling that it is derogatory to the Creator to suppose him to have created anything without a good use; on the other, it shews God to have acted, as it were, capriciously in not having informed man of the value of the thing for thousands of years; and in many instances he has doubtless not yet discovered the uses. So that in all such cases the animals, plants or minerals, are practically useless.

It is quite clear, then, that the existence of every plant or animal under their present distribution and characters cannot be solely for the good of man. He, however, like the animals, simply uses those he requires, but he has to make discoveries and experiments to find out what are most suitable for his own purposes. Indeed, nothing more can be stated as to the design of creation being for man's good, than that God has placed him in the midst of plants and animals of almost infinite diversity of form and properties, and has left him to his own intelligence to make all the use he can out of them, without any clue as to which may be useful or not, beyond what he can discover to be so for himself.

We must take a far wider view of creation than the popular one I have mentioned, though at the same time we must bear in mind that man is still the ultimate object of it, at least up to the present time.

It was necessary to enter, at least briefly, into these considerations, for the popular statement above quoted would seem to be very generally received, and expresses a supposed recognition of *absolute* beneficence as shewn in the creation of animals and plants for the use of man. This.

popular view may have been based upon the idea of direct creative fiats, or drawn from Gen. i. 29, &c.; but even if Evolution be allowed to take their place the notion would not be altered: for the supposed design does not lie in the methods of production, but in the objects produced by the will of God.

But while believing all creation to have been designed for man's use, the popular mind cannot but recognize noxious weeds and noisome beasts, and either looks upon them as objective witnesses to the curse pronounced upon the ground for man's sake or else—as we have seen—believes them to be of some but unknown use.

Now I maintain the question is a fair one to ask, Are many of use at all? One more or one less, and if one then many, excluded from the world, would make no practical difference. Plants and animals are constantly becoming exterminated, and their loss is not perceptible. But more than this, for it is not merely a question of use, there is the fact that many hurt and injure us and others. There are, for example, the carnivorous animals living upon herbivorous, and this is true for every class of animal life. And when we investigate plants we may ask, If vegetables are for food to support the herbivora, why should poisonous plants exist, or such as animals cannot eat even if not poisonous?

In believing animals and plants to have been brought into existence by the laws of Development or Evolution, we might at first imagine that it was a *necessary* result of circumstances under which certain offspring were first placed, that they developed in those 'noxious' directions accordingly, and *vice versâ*, for such as are wholesome or useful. The appearance of fixed laws is so great that some evolutionists seem not to be able to bring themselves to see the hand of Deity in Nature at all! But when Geology reveals the fact that at certain periods in the world's history—say the carboniferous—we do not know a single animal or plant he could have eaten, there does appear wisdom in introducing man in the later tertiaries when food suited to him had become abundant; notwithstanding such creatures as are necessary for his existence being associated with deadly poisonous herbs and destructive animals.

Bearing in mind, too, or *assuming* that man is the ultimate object of the design of Creation, a corresponding wisdom and forethought becomes apparent in the evolution of such beings, animal and vegetable, as should be useful to him[1].

We must never forget that the word 'use' is purely relative, and not absolute. Weeds are useless to a farmer, not to say injurious, when growing amongst his crops; but when ploughed down in a fallow field they are useful as contributing to the richness of the soil.

Just as 'Dirt is matter in the wrong place,' so weeds may be said to be 'plants not wanted at the time.' But, more than this, in consequence of Evolution myriads of beings have lived and died which could neither be useful

---

[1] The author cordially agrees with the well-known remarks of Prof. Owen as to the design in the development of the horse, exhibiting in the gradual approximation to its present structure the unfolding of the divine plan in its line of ancestry. See *Comp. Anat. and Phys. of Vertebrates.* Vol. III. p. 795.

nor injurious to man; and so too myriads do the same now; yet whenever a weed or an animal happens to fall in man's way and strikes his attention by its objectionable character, or is any way offensive to him, then does it at once so far detract from his idea of absolute perfection, and just so far has it a *disciplinary* value over him.

Indeed, all things which at any time seem to man to be objectionable or injurious are part of that grand scheme of probation, which finds illustration in every phase of man's condition upon earth.

But now, let it be particularly observed that this probationary state of things is due to the methods of Evolution; and that all 'physical evils,' from the most disastrous calamities that can befall man to the least troublesome of annoyances, meet with no explanation until the conviction of a future existence is thoroughly laid hold upon, and that the present one is in every way a disciplinary state of probation for man.

So, after all, the positive expression in Genesis as to the curse upon the ground finds a certain corroboration in nature in the evolution of every plant and animal which man may *now*, but which he would not had he not degraded himself, consider an 'evil.'

But it is to be observed of all God's disciplinary processes that their value as such is more subjective than objective, for they are disciplinary or not according to the spirit in which we look at them, not quite in proportion to the actual trouble, annoyance or pain they may cause us.

Hence we see that the evolution of noxious weeds and noisome beasts, from this human point of view, has been one of the means appointed to teach man not to expect a Paradise on earth: and that the more he frets against his condition here, the more he may feel inclined to murmur against God for allowing annoyances, troubles, and bitter afflictions to affect him: so that all external ills or evils, or whatever they may be called, appear as so many 'curses' to man in proportion as his own evil nature requires to be disciplined.

On the other hand, the sooner he learns to regard every detail of the circumstances of life as so much discipline, the sooner will he become humble and submissive to God: in the same degrees will he feel less and less the penal character of physical evils, while at the same time rebellious anger melts into filial love, a thoroughly contented and trustful spirit replaces a discontented one, and he waits for his final redemption in patience and hope.

This was the view of the great Apostle St Paul, and all that I contend for is, that such a present state of things is due to the will of God, and that He has brought it about mainly by Evolution.

I make, therefore, no apology for quoting at length a passage which so exactly represents the ideas I have endeavoured to express.

"I reckon that the sufferings of this present time are not worthy to be compared with the glory which shall be revealed in us. For the earnest expectation of the creature waiteth for the manifestation of the sons of God. *For the creature was made subject to vanity*, not willingly, but by reason of Him who hath subjected the same in hope: because the creature itself also shall be delivered from the bondage of corruption into the glorious liberty of the children of God. For we know that *the whole creation groaneth and travaileth in pain together until now.* And not only they, but ourselves also, which have the firstfruits of the Spirit, even we ourselves groan within ourselves, waiting for the adoption, to wit, the redemption of our body." Rom. viii. 18.

To the pure in heart alone does God disclose Himself and reveal His works of love. What is proof of wisdom and love in trial or discipline to the faithful believer in God, to the infidel, scoffing sceptic, and atheist, is turned into a curse[1]!

---

[1] It may be noted that what is here expressed as to the appreciation of man with regard to his condition upon earth, finds its parallel in his future estimate of things, as shewn by the parable of the Vineyard. The *actual* reward of a penny was the same to all; but its *relative* value varied from zero to infinity according to the spirit of the recipient.

To the humble mind is revealed intuitively the deep meaning of this world's *inideal*[1] condition, whereas such is used by the scoffer as a witness to the incapacity of God or else to the machinations of Satan!

But, let us remember man's moral constitution is not an absolute unchangeable thing, but admits of development or a 'growth in grace'; and let us recognise the will of God in making it such; then we shall at once see the wisdom of God in adapting the external world to correspond with, and to be of such a character as to *allow* that development in man to take place: and so the very regeneration of man falls under the same great laws of Evolution which bind all the phenomena of nature under one great harmonious principle.

On the other hand, had everything been absolutely and perfectly adapted to man's wants

---

[1] *Inideal, Inideality,* are words proposed by the writer to express the *relative state of perfection* which exists in this world. See *the Law of Inideality* more fully treated of in his paper, *Natural Theology considered with reference to Modern Philosophy,* loc. cit. p. 23.

and desires, no progress would have been possible, and other conditions than those to which he had been accustomed could be no improvement upon them: but as it is, man admits of progress, and external conditions are exactly such as to secure his moral development by means of a probationary existence.

Having seen how the final cause of all animal life finds its explanation only in that of Man, and that it is more in the sense supplied by Evolution that creatures have been of *use* to man, than that each animal and plant should be individually of some good to him; we were led on to consider somewhat of the apparent intention of God in permitting objects which seem obnoxious to us to be in the world; and we discovered that they are so more in a relative than absolute sense, and moreover that their existence is in harmony with the grand scheme of probation which embraces all terrestrial conditions in their bearing upon man.

I pass on now to notice more particularly certain features of creation, arising out of Evolution, and which have not hitherto received the candid attention by writers on natural theology, to which they are entitled, when the object is to endeavour to trace wisdom and beneficence in the designs of the Creator.

One of the most remarkable results of this mixture of 'good' and 'evil' in the world, while we regard God as the author of all, is, that at one time he seems to have permitted or created an 'evil,' at another to have created an antidote for that evil.

Thus, in allowing diseases to enter as a phenomenon of living things, He has at the same time imparted to organic nature the power of reparation from injury and recovery from disease, as well as furnished remedies which man can discover by his intelligence.

While supplying man with animals as his food, at the same time He has allowed some dozen or more parasites to attack and injure him[1].

---

[1] See the writer's strictures on a remark by Mr Herbert Spencer as to the existence of two Principles 'Good' and 'Evil,' *Natural Theology considered with reference to Modern Philosophy*, *loc. cit.* pp. 23, 24.

Again, while furnishing animals with the means of concealment, defence, or escape from their enemies, He has supplied those enemies with means of concealment too, as well as of attack and capture.

Now this observation bears upon a remarkable misapprehension in the reasoning adopted by a certain class of thinkers, who not unfrequently fancy they see the wisdom or benevolence of God in the one class of phenomena, *i.e.* when there appears to be some 'advantage' or 'good'; but do not seem to see that the 'disadvantage,' or 'evil[1],' is as much due to the Creator's will as the other[2].

This phenomenon of the existence of a mixture of 'good' and 'evil' which is of very wide application, is a result of that 'struggle for existence' which is a well-recognized law

---

[1] It must be distinctly understood that the word 'evil' as here used only signifies the popular expression for certain physical facts, and in no way implies sin or its effects.

[2] As an illustration :—It was once remarked to the writer that the presence of 'docks' (*Rumex*) was a wise provision of the Creator where *stinging nettles* abounded!

of Nature and is applicable to all kinds of life. To a partial or restricted view this law may appear as a 'fatal' result of undirected natural forces; but I maintain though God chose to establish such, as a condition of life on earth, *He had a motive in so doing:* and that motive was in its ultimate bearings to surround Man with inideal circumstances, and so render his life on earth probationary in every way.

One of the most curious of phenomena associated with this peculiar feature of organic life, mentioned in the preceding paragraphs, is what has been termed *Mimicry*.

It is now known that this principle extends over very many forms of animal and even vegetal life: not that it obtains absolutely in the structure or appearance of every one of such mimetic forms, but that there are probably few groups in which some member is not 'protected' by its external appearance assuming that of other animals with which it is usually associated, or else of something in the environing circumstances of its habitat.

But, what is most remarkable, not only may mimicry be a feature of such creatures as are preyed upon, but it is sometimes also of the animals which prey upon them; so that while the former have a fair chance of escape, the latter in their turn have a fair and like chance of capturing their prey[1].

It might be questioned how far wisdom and beneficence find place in these remarkable phenomena. A direct and favourable answer is not easy to see. Nevertheless, consistently with the 'inideal' scheme adopted by the Creator, the result—and we can only judge of motives by results—is that a fair share of happiness and immunity from care is apportioned to every animal.

That mimetic features are due to laws of Evolution will scarcely be doubted or denied: as, like all other works of design in Nature, they are not each absolute and perfect, but admit of gradations from one extreme to another; on the

---

[1] Many illustrations will be found in Mr Wallace's work, *On Natural Selection*, pp. 45 to 129.

one hand while little or no trace of mimetism is apparent in some forms of a group, yet we come at last through gradations to creatures of the same group which seem to be decidedly mimetic and designed.

It is not the purpose of this Essay to attempt to furnish theoretical views as to the secondary causes which have brought about Evolution, and therefore it will not be urged that 'natural selection,' 'the sense of sight,' or other cause, may have induced mimetic appearances. I simply recognize these appearances as curious phenomena of creation, and would observe that the relative immunity from attack on the one side, and the correlative power of capturing prey on the other, are consequences of mimetism.

CHAPTER XIII.

THE LAW OF INIDEALITY AND THE IDEA OF PERFECTION.—CONCLUSION.

It cannot be too clearly understood, nor too persistently remembered, that one great result of the process of development or Evolution, as it has been carried out in the production of animals and plants, and we may add the diversity of structure of the earth's 'crust,' is that want, or falling short, of the *ideal* or any absolute degree of perfection which the human mind can conceive. In other words there obtains universally that relative or *inideal* condition of things which is a result of Evolution.

No sound philosophical speculations or deductions can be made unless this principle be thoroughly adopted.

In addition to what has been already said on this matter, it will not be necessary to furnish more evidence than is afforded by selecting two prominent features which especially illustrate this principle.

The first to be discussed will be found in the structure of organisms, and is known as *Rudimentary organs:* the second instance of that want of perfection will be found in the conditions of the sphere of existence of all organic beings upon this earth.

Archdeacon Paley, in his *Natural Theology*, or, as he defines it, "The discovery of evidences of design attesting to the existing attributes of the Deity collected from appearances in Nature," takes anatomy as the subject of his discourse, and argues well and soundly from the structure and use of organs as to the existence of design and thence to that of an All-wise Designer.

But, he does not draw deductions from rudimentary or aborted organs, nor from useless forms of structure. Yet they are as patent to any student of anatomy as are those of full function whether of the animal or vegetal kingdom.

What argument do these rudimentary organs furnish? At first sight apparently as much against design as well-developed organs do for it!

It is true that some of the more advanced writers on Natural Theology have taken note of them, but for the most part they have been passed over in silence. Hence the argument of design, as hitherto propounded, is not guarded against one point of attack from those Evolutionists who do not recognize design in creation at all.

Dr Whewell's interpretation of the meaning of rudimentary organs will illustrate the only idea which could be brought forward at a time when Evolution was not generally accepted.

That learned writer says in his *Plurality of Worlds*, p. 345:—"In the plan of creation we have a profusion of examples where similar visible structures do not answer a similar purpose,— where, so far as we can see, the structure answers no purpose in many cases, but exists, as we may say, for the sake of similarity, the similarity being a general law, the result, it would seem, of a creative energy, which is wider in its operation than the particular purpose."

This is clearly no explanation at all, and we almost feel inclined to adopt Mr Lewes' expression, that such reasoning is "a specimen of pedantic trifling worthy of no intellect above the Pongos[1]."

Now, the doctrine of Evolution satisfactorily accounts for the existence of rudimentary organs, and no other theory has as yet done so.

The writer must assume the readers of this Essay to be more or less familiar with instances of rudimentary organs, and does not, therefore, deem it necessary to bring forward more than may be required for the general argument, in support of the wisdom of the Creator, as displayed in the evolution of animals and plants.

Their importance in regard to any discussion of this sort cannot be overrated, for they have a

[1] *Mr Darwin's Hypotheses*, by Mr G. H. Lewes, in April, June and July Nos. of the *Fortnightly Review*, 1868.

strong *primâ facie* evidence against such an attribute, and one which opponents have not failed to seize upon as witnessing—from their point of view—to the working of *immanent* forces instead of a personal God[1].

On the other hand, the writer maintains they really testify to the harmony or unity of those fundamental principles of Evolution, upon which the scheme of creation has been founded by the Creator.

Now, the first inference deduced from a careful study of *all* kinds of organs is that while design appears very pronounced in some, it is wanting altogether in others, such as rudimentary structures. Secondly, of the very best instances of design—such as the eye—no organ can be called *absolutely* perfect, nor one than which we cannot conceive a better.

Thus, as the writer has elsewhere expressed[2],

---

[1] *E.g.* the following are Mr Lewes' expressions: "bungling," "Nature feeling her way," "tentative," "blundering," &c. *loc. cit.*

[2] *Natural Theology, considered with reference to Modern Philosophy* (*Transactions of the Victoria Institute*, Vol. VII.).

from such an elaborate organ as the human eye to a mere pigment-spot of an Echinoderm—from the well-developed legs of the majority of Lizards to the rudimentary and useless representatives of legs in certain snake-like genera—organs of varying degrees of character can be found which impress us proportionately with corresponding degrees of evidence of design: so that as structures apparently adapted for certain ends in some organisms are found less and less so in kindred forms, so design, as applied to the former, from being very pronounced becomes, as it were, less and less so until it disappears altogether. Thus if the following genera be compared, it will be seen how a gradual degeneration of the limbs (both in the legs and the number of toes) indicates, so to say, a corresponding dying out of purpose, till at last nothing remains but rudiments under the skin, in which the intention of locomotion is finally gone and that design has disappeared altogether:—*Zonurus griseus, Tachydromus sexlineatus, Saurophis tetradactylus, Chamæsaura anguina, Pseudopus Pallasii.* Now these examples are isolated instances in as many distinct *contemporary* genera. The same phenomenon may be witnessed in successive generations from an early type. Thus the *Palæotherium* had three well-developed toes on each foot, the central one being slightly the larger. In the *Hipparion* of a later period the two lateral ones became much smaller, and nearly resembled the pair of rudimentary toes of a cow, while the central toe and its supporting bones were proportionally larger. In the present epoch we have its descendant, the horse, with only one toe or hoof, nothing but the rudimentary 'splint bones' corresponding to the two lateral lost toes remaining. Nature is replete with such illustrations of rudiments, and the tertiary strata at least abound, as we have seen, with these transitional forms, in which organs in the earlier cases possessed functional power, but gradually disappear in the later and are finally lost altogether.

Hence we see that while on the one hand innumerable examples can be found such as

Teleologists have hitherto seized upon for their illustrations, and which to a believer in a personal and creating God evince unmistakeable and admirable design; on the other hand a large class of structures can be pointed out which either scarcely admits of the word at all, or else seems to militate against it entirely.

The explanation, then, hitherto offered by Natural Theologians—such as that by Dr Whewell—is quite inadequate, not to say unphilosophical, and directly opposed to the very principles upon which the argument of design is based.

On the other hand, these rudimentary structures cannot be accounted for except by Evolution; to which doctrine, however, they form one of the strongest witnesses. For, were any abrupt changes of structure constantly occurring, we should at once begin to infer that some power was as constantly at work to interfere and to make such changes, somewhat after the notions of the cataclysms and recreations of the early geological theorists. When sudden breaks between allied forms appear to occur, the balance of probability is greatly in favour of the inference of the previous existence of extinct forms, which once united such well-differentiated types as may now exist[1].

It may be objected that rudimentary forms are sometimes regarded too much in the light of atrophied conditions and not as origins of future organs; and it is worth while observing that there are two ways of regarding them, both, however, equally in accordance with the doctrine of Evolution; and in many cases it is at present impossible to say with certainty which would be the correct view. Thus, in the case of the lizards, it may be that the condition

---

[1] The writer is quite aware that many anatomists still cling to the idea of an 'absolute use,' and say that rudimentary organs *have* such though it is not always to be detected, because some may appear to be resolved into excretory organs or as points of attachment of muscles, &c.; but a thoroughly impartial view, the author believes, would dispel such notions. Without denying such uses as the above being possible, he yet maintains that rudimentary structures exist which *are* useless and could be dispensed with without the slightest detriment to the individual; such as the rudimentary pistils in the male flowers of such as are sexually distinct.

of the limbs in *Pseudopus*, which are rudimentary and concealed beneath the skin, was the forerunner of the state of the limbs represented by the other genera given above. The argument, however, is equally sound on either supposition. On the first, the design of the limbs dies out, and is replaced by the snake-like method of progression; on the other, the latter mode of locomotion gradually disappears and is replaced by limbs.

As another instance we may compare the rudimentary wings of a Penguin with those of the Struthious birds. In these latter we seem to have an affinity to the *Dinosauria*, so that there is some presumptive evidence in favour of the wings of the *Struthionidæ* representing the antecedent but arrested condition of those with full-power, as of the flying *Carinatæ*. On the other hand, as the Penguin belongs to this group, its wing might reasonably be regarded as an atrophied condition, but modified, subject to *the Law of Compensation*, to enable that bird to dive and swim.

The law here mentioned is one of great importance, and disproves the justice or reasonableness of any charges, founded upon the existence of rudimentary organs, which the Positivist or Materialist may make.

This law is of universal application, and fully discovers to us the power of the Creator. For, if it were asserted that it was proof of feebleness to allow an organ to degenerate and its function to vanish, it is as much a proof of power to be able to compensate the organism for such loss, by extra development in some new direction. The instances given above will suffice to illustrate this principle: and the wisdom of such a method is obvious in thus rendering creatures capable of great variety of form and function, which would have been much restricted with other and more uniform structures; and moreover enabling a greater diversity of life to maintain itself under the ever-varying conditions of environment.

It is a law or principle which is found ranging over the vegetal kingdom as well as the animal. A few illustrations will shew this. Weak stems are often compensated for by powers of climbing; so that more plants can grow on a limited area for the most part covered with trees and shrubs, as in the Brazilian forests. To effect this process of climbing various are the methods. In 'stem-climbers,' such as the hop or bindweed, the stem bows in all directions until it has swept round some support. In *Clematis* the leaf-stalks are highly sensitive to the touch, and clasp any object they may meet. In the vine the flowering branches sacrifice their power of producing buds to become highly sensitive tendrils. Again, in the *Compositæ* many of the 'florets,' as in the cornflower, sacrifice their 'essential organs' for the enlargement of their corollas, probably in order to render them more attractive to insects; and so are better able to secure 'the intercrossing of distinct flowers' in order to avoid too close interbreeding.

It will probably be at once noticed that this law is after all but a general illustration of that great principle of Nature already discussed, which may be called 'the continual effort of beings to arrive at mutual and beneficial adjustments.'

Again, when we recognize, in the vegetal and animal kingdoms alike, the seemingly universal tendency to separate the sexes, we may ask why do they ever both occur in the same flower or the same animal? If this question has no other answer than, such is the will of God, it is at least satisfactorily accounted for by the theory of Evolution of beings from the more generalized types or grades to the more specialized: so that this differentiation of the sexes is simply part of that very process, hermaphroditism therefore representing the undifferentiated state, and wherever it still exists, pointing to arrested conditions.

The second instance of Nature falling short of that absolute degree of perfection, which we can conceive possible, is to be found in the conditions of the sphere of existence of all organisms upon the earth.

This has been expressed in some such words as the following:—"Animals and plants do not live where circumstances may be best suited to them, but where they *can* or where other animals and plants will respectively let them live."

Hence, the intense struggle for existence, which obtains everywhere in the constant attempt at self-adjustment to external circumstances, can never effect more than an *inideal* or relative state of perfection.

Conditions of happiness or prosperity circulate about an average or mean, but never attain anything approximating to an absolute state of perfect harmony for every day of a creature's existence.

But a more important fact will be seen to arise out of this: and that is, not merely does a certain degree of unsatisfactoriness exist in the conditions of life, but such an amount of derangement may occur, as to be regarded in the light of the so-called physical evils generally; this may otherwise be expressed by saying that there is a continual process of an attempting to secure a mutual adjustment between the three great kingdoms of Nature, which is never absolutely attained; and as a concomitant effect of such attempts, is the continual clashing of opposing forces, so that while only a partially satisfactory adjustment can be secured, it does not obtain without much imperfection and often great suffering.

Thus *e.g.* earthquakes, volcanos and deluges are physical evils to which Man is subject, because he has appeared upon this earth before the equilibrium of Nature's internal and external forces has been secured. In other words, there is a constantly prevailing want of adjustment, or statical conditions, between those forces of Nature which would secure to organic beings an immunity from all chance of destruction. Similarly with aërial phenomena; rain is necessary for animals and plants, yet the due supply ever ranges round a mean. In some years there is not enough for the crops: in others, they are irretrievably damaged by excess. Again, the fact that half the animal creation preys upon the other half, is often regarded as an 'evil' (whether justly so is another question, and one to which the writer would unhesitatingly reply in the negative). So again, many diseases to which plants, animals and Man are alike subject, are due to abnormal combinations of the ultimate elements of the beings attacked; and arise out of a want of a perfect system of combinations of those ultimate elements, which would not admit of any disturbance whatever, and which we can conceive possible, but which is probably never secured; for, to be absolutely perfect, our bodies should not be liable to disease at all.

Now, in considering Man as part only of the scheme of creation, it will be desirable to compare the effect of physical evils upon animals with that upon Man. We have no reason for supposing animals to be conscious of any of these things at all: nor do physical evils trouble their minds. They are in no perplexity as to their meaning. They live their life and die, and there is an end of them. It is not so, however, with Man. He alone can reflect upon the conditions of his existence; and he finds from the most impartial scrutiny of his surroundings that they give rise to no higher conclusion than that he is in the midst of a mixture of good and evil, or a relative state of happiness; and, as he learns from Revelation, a state of probation or trial, in order to test, to experience and to qualify him for a future existence, without a belief in which this world has ever seemed to be dark and gloomy.

Reflecting therefore in this way, we find Man's future state to be the sole interpretation not only of his existence here, but, through him, of all terrestrial creations as well; and the probationary condition under which Man lives, while on earth, is secured by that relatively perfect or inideal state of things which universally obtains: and lastly, that very *inideality* is essentially the correlative result of the processes of development and evolution.

Thus, then, do we see all life linked together: and as far as we can see the wisdom of probation as anticipatory of a future state of happiness, so far do we recognize the wisdom of the plan adopted and carried out by the Creator in bringing about the changes of the world, animal vegetal and mineral, which by their combined efforts have evolved the present conditions of things and rendered them fit for man's terrestrial existence.

We do not forget that God could not only have made Man at once fitted for heaven, but have placed him there; so that the wisdom of development and the wisdom of probation are alike synonymous with the will of God. They are the necessary correlatives of that system of progress which has a universal application in every class of phenomena that admits of improvement or differentiation of any kind.

Indeed, we may go one step further and say, that were this not the case heavenly conditions could have no difference from earthly ones; or rather, as they might of course be different, that under the present limitations of our existence we cannot conceive of any method by which heavenly conditions should be made contingent upon earthly ones. So that, *The whole scheme which God has framed for Man's existence, from the first that was created to all eternity, collapses if the great law of Evolution be suppressed!*

Let us clearly understand, then, that it was the Will of God that the environments and conditions of life should be only relatively perfect in this world, and that the chance of physical evils injuring or destroying beings endowed with life appears to have been predetermined, is an integral part of those conditions, and can never be removed before the end of the world shall come!

Recognizing this, we will briefly trace the effects of physical evils upon creatures of successive grades or types. In plants and the lower animals, such as sponges, there is no consciousness of pain, and therefore these evils are inappreciable. With other animals, as we ascend from the lowest to the highest, we find elements gradually differentiated by which the creature can become conscious of pain, bodily at first only. But in the higher animals *fear* and *anxiety* are superadded, and we get the germ of mental anguish; but as yet no consciousness of what physical evils *are*. As soon, however, as we step across the barrier we find Man, not only subject to physical and mental pain of every degree, but *conscious* of his surroundings being imperfect, able to reflect that he has acquired a "knowledge of good and evil"; and this new power makes him suffer mentally by *anticipating* evils, and these expectations of coming evil are often far more exquisite in their mental torture than the actual endurance of the evils themselves.

It is the existence of physical evil which has ever perplexed the thoughtful man in all ages; and until the light of Revelation was thrown upon this world's mysteries it is not to be wondered at that the best antidote to misery was expressed by—" Let us eat, drink and be merry, for to-morrow we die." But, it is just because a future life is offered that we see why "we live in jeopardy every hour": and why that very yearning for a happier state, or at least for *rest*, should be so strong and so deeply seated in the human heart.

In drawing this Essay to a close, it will be noticed by the reader that so much has been said upon the relatively perfect or inideal conditions as well as the physical evils of the world, that he may almost imagine that the writer could not see any signs of the goodness and beneficence of the Creator. This would be a false, not to add an unjust, conclusion. The fact is that so little has hitherto been said upon physical evils by writers on Natural Theology, and so one-sided, therefore, have been the general observations about the wisdom and beneficence of God, that the present writer deemed it desirable to bring forward into greater prominence the recognition of physical evils of the world as part, and a very important part, of the scheme of Creation as bearing upon the probationary condition of man.

That animals, as a whole, enjoy their existence cannot be doubted: it may be of a limited character, but it is proportional to their powers of enjoyment; it may be regulated by law, but still their minds are not capable of any original thought; and so, there is no appreciation of the limitations to which they are subject; they seem to have no fear unless an enemy approach them: so that considering that it was solely the will of God that they should exist at all, we may safely say that He has showered upon them in their wild state a condition of happiness, an enjoyment large in proportion to their relatively perfect environing conditions.

When, however, we consider Man, our own personal experience supplies us with a gauge of the amount of happiness we enjoy. Suffice it to say that both philosophy and religion teach us that happiness is, to a very considerable extent, left to our own powers. For, as has been already remarked, just as external evils are mainly subjective, *i.e.* they are evils or not, according as we ourselves estimate them; so, where one man passes a life of "godliness with contentment," another sees nothing but troubles, difficulties and anxieties. This power of acquiring happiness, even when surrounded by difficulties, is an undoubted proof of beneficence in God, Who so ordained that such should be the enduring reward of faith in Him.

It may seem strange first to have permitted if not created an 'evil,' and then to introduce a 'good' act to counteract it: but, then, such is the principle upon which God has ordered this world, and is in fact part of that disciplinary process which is the final cause of inideality.

No other words will better explain Man's estate than "Seek ye first the Kingdom of God and His righteousness, and all other things shall be added unto you." For such is God's providence; and although we cannot expect absolutely perfect conditions in this world, nor does a godly life necessarily ensure a worldly fortune, yet we do practically find that if we endeavour to live godly, righteously and soberly, we are securing just those conditions which are most likely to ensure such worldly prosperity as is good for us.

God's providence is not to be questioned because a worldly fortune may not be ours. Our conditions of happiness are not to be measured by pounds, shillings and pence.

If there were no future life, there might be some show of reason for setting high value on temporal prosperity; but on the great principle of future life being made contingent upon our right use of a terrestrial state, no such value ought to be assigned to any temporal conditions of satisfaction whatever.

But in the midst of this disciplinary process,

who, if he possess a particle of Faith, will deny the beneficence in His providence, or the goodness of Him from Whom cometh all good and perfect gifts? Who in His love showers down His rain upon the just and the unjust, Who, in His infinite wisdom and goodness, has made the law that the giver should find more satisfaction than the recipient, and so prepared the heart for charity, and ruled that the man who sacrifices himself should know of His love; that he who is content with his lot—believing all things to work for the good of those who love Him—finds that rest and satisfaction in contentment to be his which the restless and dissatisfied yearn after and can never find; yet is it already prepared for them of God, and within reach of their very grasp!

Who, I ask, that has once 'tasted of the goodness of the Lord' will deny that He is a Father of Love to his children, even though He try them as by a fiery ordeal or suffer them to pass through the furnace of affliction?

www.ingramcontent.com/pod-product-compliance
Lightning Source LLC
Chambersburg PA
CBHW021815230426
43669CB00008B/759